Lean Six Sigma

Leadership Tools for Black Belts

Lean Six Sigma

Leadership Tools for Black Belts

Mary McShane-Vaughn, Ph.D.

UNIVERSITY
TRAINING
PARTNERS

Six Sigma University Press

Six Sigma University Press

Folsom, CA

https://courses.6sigma.university/pages/books

Library of Congress Cataloging-in-Publication Data

McShane-Vaughn, Mary

   Lean six sigma leadership/Mary McShane-Vaughn, Ph.D.

      p.cm.

ISBN-13: 978-0-9906838-8-9 (University Training Partners)

   1.   Quality Control – Statistical methods – Handbooks, manuals, etc.

I. McShane-Vaughn, Mary II. Title

For my daughter Kate, who personifies two lessons that we have tried to instill in all our children:

1) Ask questions when you don't understand something.
2) "Smart" will only get you so far. To succeed, you must roll up your sleeves and do the hard work.

Thank you for asking me all those AP Calculus questions as I was writing, editing, and formatting the first edition of this book. Our nightly study sessions were a welcome respite from fighting with DPI issues and the intricacies of odd and even header formats.

Table of Contents

Preface

This book is written primarily as a companion text for the first course in an online Lean Six Sigma Black Belt program created and delivered by University Training Partners. However, it can be equally as useful to any quality practitioner who wants to add more tools to his or her Six Sigma toolbox, or who is preparing for a certification exam.

The text covers non-statistical skills in the American Society for Quality (ASQ) Black Belt body of knowledge – skills that are vital for leading successful Six Sigma projects.

The truth is that the hardest part of leading Six Sigma teams is not the data collection or statistical analysis: it's dealing with people. The "soft skills" of project selection, project prioritization, project scoping, stakeholder analysis, team building, voice of the customer, training, and project management all are employee or customer-facing – and difficult. As I tell my students, you can always hire a contract statistician to do the heavy lifting for you if you run into a complex analysis situation, but you cannot outsource managing your people.

This course companion text includes all the printed materials students will need for the Lean Six Sigma Leadership Tools course and will be used in conjunction with the course module presentations, companion spreadsheet, and course assignments. In the appendix is the resort hotel case study that is used throughout the course to allow students to apply the tools they learn.

This text will also be an invaluable resource for the online course quizzes. Feel free to tab the book and write in its margins. Enjoy!

To order more copies of this book, or to learn more about other Six Sigma and statistics courses University Training Partners has to offer, please visit our website: https://courses.6sigma.university

Acknowledgments

I would like to thank my husband Jim for taking time out from his busy consulting schedule to perform a detailed read-through and edit of the material in the manuscript. His careful copyediting and thoughtful suggestions have made this text a much more readable and informative course manual. His expertise in the behind-the-scenes functionality of Microsoft Word was a lifesaver, as was his patient coaching on the tedious process of preparing the manuscript for publication.

Chapter 1 Getting Started

This text provides the student with material to accompany the Lean Six Sigma Leadership Tools course developed by University Training Partners. The course is the first in a three-course sequence designed to provide candidates with a thorough presentation of Lean Six Sigma Black Belt methodology and tools. Students will use this text as a companion to the online materials, discussions, assignments, Microsoft Excel exercises, and quizzes presented in the course.

This text, and the Lean Six Sigma Leadership Tools course itself, covers the non-statistical topics that Black Belts need to launch and lead successful process improvement teams. Materials in the chapters align with the DMAIC project phases, as well as the online course modules, and include: Reviewing Six Sigma; Launching Projects; Capturing the Voice of the Customer; Choosing Metrics; Leading Teams; Managing Projects; Mapping & Graphing Data; Finding Root Causes; Implementing Lean; Calculating Lean Metrics; Training Employees; Designing for Six Sigma; and Closing the Project.

In the first week of the course, students should familiarize themselves with the online course layout and tools. Students are encouraged to read through the Mirasol Resort Hotel case study located in Appendix A this first week. Many examples and exercises will be based on the case company, a boutique hotel facing numerous challenges.

For students serious about pursuing a career in the quality field and functioning as a Black Belt in industry, I suggest they join the American Society for Quality (ASQ). Members enjoy access to many white papers and case studies, a monthly print magazine, and discussion boards. Members also receive discounts on quality references and textbooks.

There are many ASQ divisions to choose from depending on your industry and interests, including one for Statistics, Lean Six Sigma, Chemical Processing, Healthcare, Software Quality, Quality Management and many more. Each of these divisions hosts a small conference each year. In addition, the ASQ World Conference is held in May of each year, with over 2000 quality professionals in attendance. Find out more by visiting www.asq.org.

Appendix B presents the 2015 ASQ Six Sigma Black Belt Body of Knowledge. While there is no governing body for Six Sigma certification, ASQ is thought by many to be the gold standard.

As you progress through the Lean Six Sigma Black Belt program, you can check off the topics on the ASQ Black Belt listing. Students can use the *Certified Six Sigma Black Belt Primer* by Womack, or the *Six Sigma Black Belt Handbook* by Kubiak and Benbow to study selected topics in more depth. (See Bibliography for details.) Pay attention to the Bloom's level of cognition associated with each topic to make sure your knowledge matches the depth of understanding required.

The online program is not meant to be a prep course for ASQ Black Belt certification. Rather, the leadership tools course presents topics and templates that are most useful for implementing and leading real-world projects in the workplace. The statistics course which follows the Leadership course will give Black Belt candidates an in-depth understanding of statistical techniques. Performing statistical analysis in the real world requires more from a Black Belt than simply answering multiple-choice questions on a certification exam. In the third course of the Black Belt program, students will identify and lead a Six Sigma project at their workplace. The three-course sequence will be excellent preparation if students do decide to pursue ASQ certification.

What's in it for me?

- Take a few moments to think about what you would like to gain from this class (WiiFM).
- Write your thoughts down on this page (we won't share these).

We will return to your WiiFM page at the end of the last class session.

(Note to students: There is no requirement to post this in the online course. You will revisit this page in the last module of the course.)

Chapter 2 Reviewing Six Sigma

The concepts and vocabulary in this chapter will provide a review of the major tenets of Lean Six Sigma you learned in your Green Belt program. Read this chapter twice. Reading the first time will dust off those cobwebs. The second time around, try reading through the chapter as a future Black Belt who will be tasked with explaining Six Sigma to various audiences. Are there any phrases that you can pull out and make your own? What types of examples would work especially well in your workplace, and which will not? Use the material in this introductory presentation to craft your personal Six Sigma elevator speech and develop your own set of illustrative examples to use in your role as a Black Belt.

2.1   The DMAIC Roadmap

The systematic approach of the DMAIC cycle, shown in Figure 2.1, is at the heart of Six Sigma. Recall that DMAIC is an acronym that stands for Define, Measure, Analyze, Improve and Control. Each Six Sigma project goes through these five steps, with no skipping. DMAIC provides a roadmap for your Six Sigma project plan.

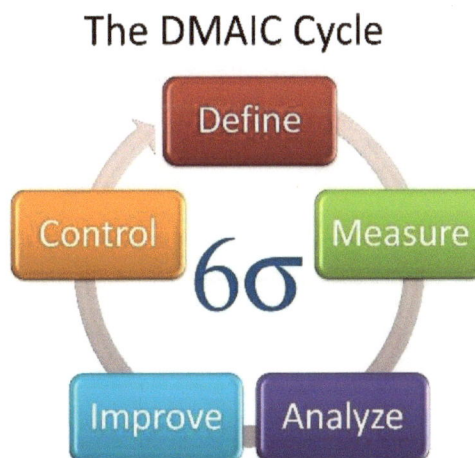

Figure 2.1 The DMAIC cycle

Many of you may have also heard of Deming's PDCA cycle, which stands for Plan, Do, Check, Act. It is also referred to as PDSA, or Plan, Do, Study, Act. The PDCA cycle can be thought of as a wheel that keeps turning. After Act, the cycle starts again with planning for the next improvement. PDCA never ends. In contrast, the DMAIC cycle shows the distinct start and end points of a project. After the Control step, the process is turned back over to the process owner. PDCA is more aligned with Lean, or continuous improvement. We will see that both approaches have an important part to play in improving quality.

Figure 2.2 (George, 2004) lists the types of tasks a team might perform in each of the DMAIC steps. In Define, the steering committee selects and assigns the project. The team leader creates the project charter with the guidance of the project champion. At this point, the scope of the project is validated. Six Sigma projects are short-term, anywhere from two to nine months, so the Black Belt needs to scope the project carefully. Project stakeholders are then identified, and from this listing a team is selected. The team identifies customers and collects voice of the customer data. Although the project charter will list dates that each of the DMAIC phases will be complete, a more detailed plan – such as a Gantt chart – is created.

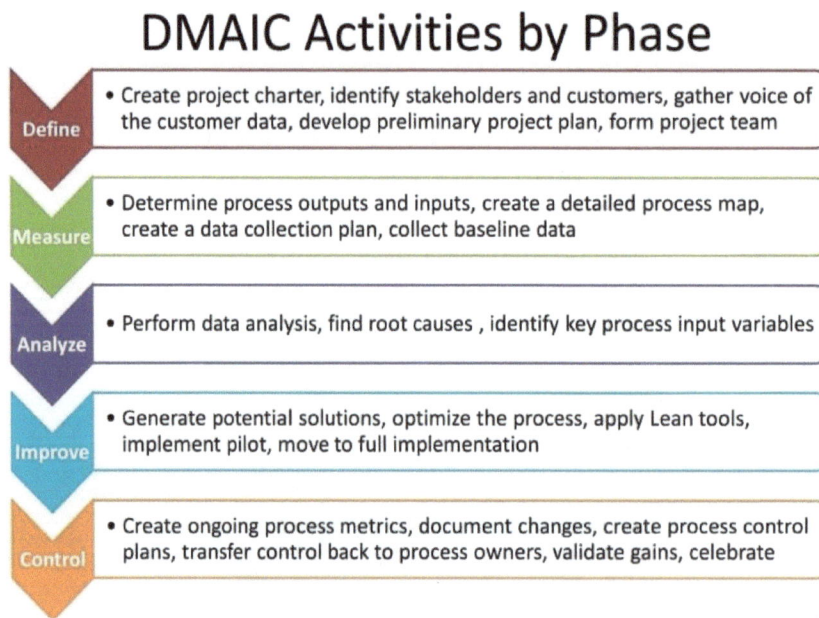

DMAIC Activities by Phase

Define
- Create project charter, identify stakeholders and customers, gather voice of the customer data, develop preliminary project plan, form project team

Measure
- Determine process outputs and inputs, create a detailed process map, create a data collection plan, collect baseline data

Analyze
- Perform data analysis, find root causes , identify key process input variables

Improve
- Generate potential solutions, optimize the process, apply Lean tools, implement pilot, move to full implementation

Control
- Create ongoing process metrics, document changes, create process control plans, transfer control back to process owners, validate gains, celebrate

3

Figure 2.2 DMAIC activities by phase

In the Measure phase of the project, the team determines the process inputs and outputs and creates a detailed process map. The process map is one of the most important tools used in any project. If applicable, the measurement system is validated at this point, and a data collection plan is created: what variables will be measured, by whom, with what instrument, and at what frequency, for example. The baseline capability of the process and sigma level can be determined in Measure as well. This information can then be used to update the project charter goal statements.

In the Analyze phase, the team analyzes the data collected in Measure and determines the critical input variables – the ones that affect the output of interest. We can refer to this as determining the relationship $Y = f(X)$, where Y is the critical output and $f(X)$ is a function of the important input variables. In this phase, data and process analyses are performed. Based on the analysis, root causes can be identified, prioritized, and tested.

Next comes the Improve phase, which is the phase that all new teams want to start with. And why not? It is the crux of the project: solving the problem. But if we leap to Improve without the DMA steps, we may be addressing a symptom and not a cause, or using gut feel instead of data for our decisions. Going down this road will result in a sub-optimized solution. After the DMA work has been accomplished, we can generate and prioritize potential solutions to the root causes found. It is in this stage where Lean principle and tools are most applicable. We can perform a risk assessment and pilot our various solutions before rolling them out full-scale.

Once the solution has been rolled out, the team moves on to the last stage, Control. Here the team prepares to hand over the process back to the process owners. New standard operating procedures (SOPs) are documented to assure changes are codified. Ongoing process metrics are established and tracked using statistical process control (SPC) tools, and a control plan is created to specify the actions taken when the process strays from its new improved level. The team documents lessons learned from the project and shares them across the company so that future teams can leverage the success of this project and avoid pitfalls the team encountered. The team also validates the gains of the project, determining the true cost-savings of their efforts. Transition of the process is then complete, and a celebration is on order.

What types of tools are used in each phase? In Define, teams or team leaders may use project selection, project charters, a SIPOC diagram, voice of the customer, Kano analysis, stakeholder analysis, Gantt charts, principles of team dynamics and quality costs.

In Measure, teams may identify types of data, create data collection plans, and check sheets, and perform measurement system analysis. In this phase, they also may create process maps, calculate baseline defects per million opportunities (dpmo) and process capability, and create a current state value stream map.

In Analyze, we concentrate on finding root causes. A variety of graphical tools help us make sense of the data we collect in the measure phase, including Pareto charts, run charts, scatter plots, histograms, spaghetti diagrams, check sheets, value stream mapping, measles charts, and box and whisker plots. Calculations such as process capability analysis, dpmo and sigma level, descriptive statistics, and z statistics may also be performed. In addition, this is the phase in which statistical analysis can be performed, such as hypothesis tests, regression modeling, and contingency table analysis. Teams concentrate on finding root causes by using fishbone diagrams and the 5 Whys, for example. Note that not all tools will be applicable or necessary. The Black Belt must know the tools well enough to select those that match the situation. There is no rule that says at least five tools must be used, or that certain tools must always be used. There are no Six Sigma police. Use what fits.

In Improve, teams may implement such tools as mistake proofing, 5S, SMED (which stands for single minute exchange of dies, also called quick changeover), Kaizen events, statistically designed experiments to optimize processes, or pilot testing.

In the Control phase, teams can implement statistical process control charting, and create standard operating procedures (SOPs) and process control plans.

How many of these tools are you already familiar with?

Now that we have gotten through some preliminary definitions, here is a quick quiz:

Six Sigma is a

a) management philosophy
b) systematic approach to problem solving
c) statistical standard of quality
d) all the above.

How did you do?

If you had answered a) management philosophy in the quiz, you would have been correct. Six Sigma, like Total Quality Management before it, is a management philosophy. It is characterized by organizational wide deployment and involvement and is driven from the top down. This contrasts with, say, Quality Circles, in which quality is a grass roots type of implementation.

In Six Sigma, there is an elevated standard of excellence, and quality is linked to the business' bottom line. Dollar savings are a key component of every Six Sigma project, which is a bit different from other quality movements.

Six Sigma requires in depth training of quality tools – think about all the time you put into becoming Green Belts, and now, are signing up for two additional classes to learn more quality and statistical tools! That is a big training and learning commitment.

Finally, Six Sigma is oriented toward projects. This is different from the Lean "continuous or incremental improvement" approach. In Six Sigma, we identify a problem, create a project, follow DMAIC for 2 to 9 months and we are done. Money saved, process turned back to process owners, and then on to the next project.

Back to the quiz. If you had answered that Six Sigma was a systematic approach to problem solving, you also would be correct (you can see how this is going).

Six Sigma concentrates on processes, and improvement projects follow the DMAIC steps. Decisions are made based on data, not by whomever has the loudest voice in the room, or the highest seniority or most intimidating personality. Think for a minute about your workplace: how are decisions made? What are the driving forces behind them? The goal of Six Sigma is to decreases variation, which translates into more consistent process inputs and outputs, and to decrease defects. Once we accomplish this, customer satisfaction will increase, and we'll achieve better bottom-line results.

We've established that Six Sigma is project driven. What is the rationale behind project identification? We have customer requirements on the one hand, which we gather through voice of the customer (VOC), as well as our business strategy. These two drivers inform the company strategic goals, and from the strategic goals come the Six Sigma projects. The steering committee will choose projects that then loop back to the customer requirements and business strategy, making a circle as shown in Figure 2.3.

The Six Sigma Project Method

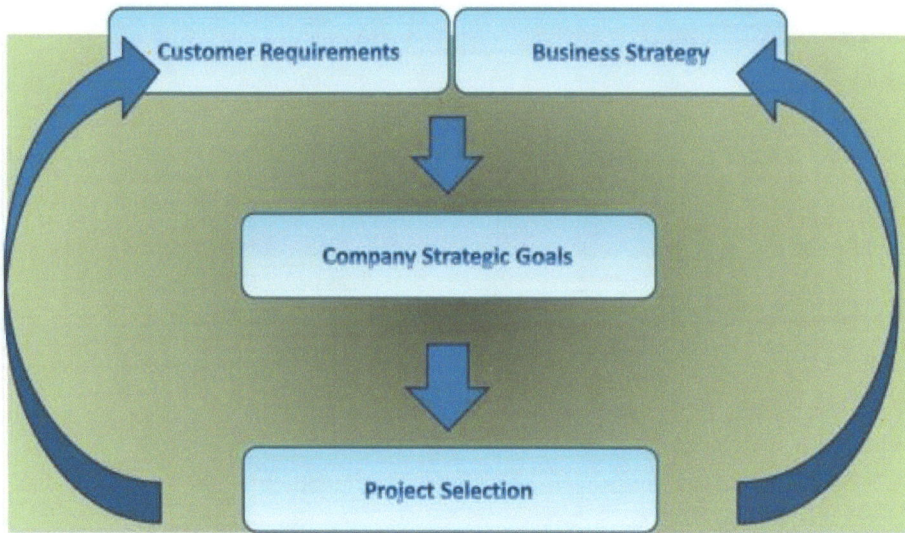

Figure 2.3 Six Sigma project method

Returning to the quiz: Six Sigma is also a statistical standard of quality, isn't it? The Six Sigma standard is defined as 3.4 defects per million opportunities. We will cover the origin of this figure a little later in the course. (On the other hand, the goal of Lean is perfection, or zero dpmo).

We will use a fast-food example to illustrate the meaning and implications of a 6 Sigma process. A major hamburger chain served 1.26 billion hamburgers monthly. Mind boggling, no? We can arbitrarily choose a defect type that a hamburger could have, say, undercooked meat. If the chain were running a commendable 99% quality level – we would all like to be at that level, right? – it would be serving 12,600,000 undercooked hamburgers! How many of those burgers would get customers sick? E coli, listeria, there are lots of unfortunate scenarios that could play out. How many customers would get so sick that they would be hospitalized, or even die? Clearly, serving 12.6 million defective burgers is a problem. Once we blow up the scale of the numbers, 99% doesn't look that good anymore!

Now, contrast this to a 6 Sigma level of quality, in which only 3.4 defectives are found per million. This translates to a 99.99966% quality level and results in 4,284 undercooked hamburgers. Still not perfect, but orders of magnitude better than our last example. That is what we are shooting for in 6 Sigma: to take that 12.6 million down to 4300. I think this is a powerful illustration.

Here are more examples showing the difference between 99% and 6 Sigma Quality using real numbers and hypothetical defects. As shown in Figure 2.4, based on the number of registered vehicles in the United States, 99% quality would result in 7011 vehicle breakdowns per day, versus 870 breakdowns per *year* at a 6 Sigma quality level. It is the difference between 2441 boating distress calls per week and 10 distress calls per *year*. In hospitals, at 99% quality, 3450 babies would get dropped each month, versus 63 dropped per year. Or the difference between 1452 surgical mistakes per day versus 180 surgical mistakes per year.

99% Quality = 3.8 sigma	99.99966% Quality = 6 sigma
• 7,011 vehicle breakdowns per **day**	• 870 breakdowns per **year**
• 2,441 boating distress calls per **month**	• 10 distress calls per **year**
• 3,540 dropped babies per **week**	• 63 dropped babies per **year**
• 1,452 surgical mistakes per **day**	• 180 surgical mistakes per **year**

With 1.5 sigma shift

Figure 2.4 More comparisons between 99% and Six Sigma quality levels

These examples are all hypothetical of course, but they illustrate the point well. The vehicle breakdown example is based on the actual the number of registered vehicles in the US. A plausible defect type was selected, and then the numbers were crunched. The same procedure was used with the number of boats, number of babies born per year, and number of surgeries per year.

Next, we examine some actual examples. On the graph in Figure 2.5, we have defects per million opportunities on the x axis, and sigma level on the y axis. Flight safety is running a bit higher than a 6 Sigma level, meaning that there are fewer than 3.4 incidents per million flights. Baggage handling runs at about a 4.2 sigma. The number of dropped cellphone calls, and incorrect advice given by the IRS helpline are about equal at 3 Sigma, or roughly 67,000 defects per million. This isn't unusual; the typical company in the US is running at about a 3 Sigma level of quality.

On the other end of the spectrum is the on-time arrival rate for airlines. That metric is running at about a 2.2 sigma level, or 225,000 defects per million. Note that going from a 2.2 to a 3.2 Sigma level shoots us across the x-axis from 225,000 all the way down to about 50,000. Great gains are made when we reduce sigma at these levels. But, after that, there are diminishing returns: the defect reduction from 3 to 4 Sigma is about 40,000, and less and less as our sigma improves. It is much easier to improve a broken process, and more gains can be made. We can think of problems as excellent opportunities. Once we start improving more, it is harder to make huge gains. Still, we would rather have no babies dropped versus 63 a year, so it is worth our while to continuing to improve.

A Few Actual Sigma Levels

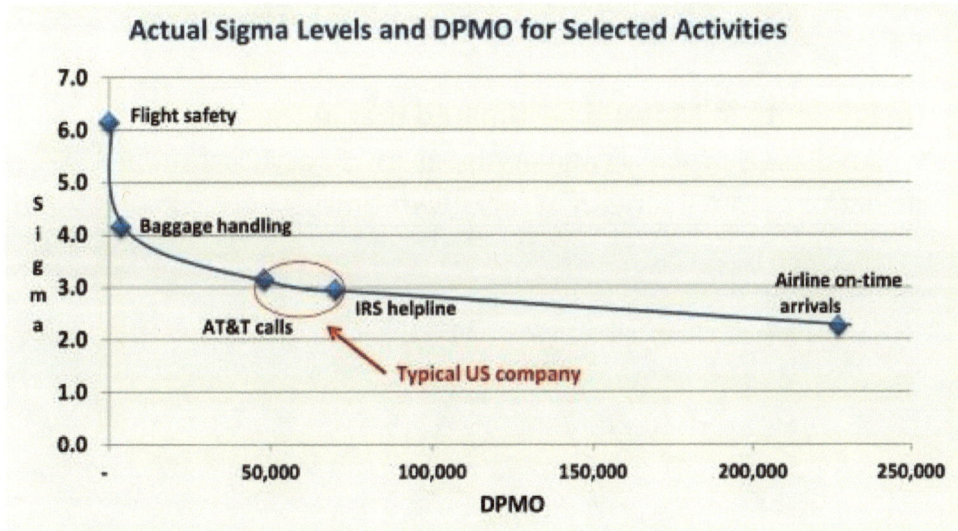

Actual Sigma Levels and DPMO for Selected Activities

With 1.5 sigma shift

18

Figure 2.5 Actual Sigma Levels

2.2 The Six Sigma Shift

In the previous examples, there was a caveat in the bottom corner of the figures stating that the calculations were made "with the 1.5 sigma shift." Where did the "shift" come from? Bill Smith at Motorola developed the concept, so we have him to thank or blame for it, depending on one's perspective. The general idea makes sense: even a well-behaved process is going to move its mean slightly back and forth over time, as illustrated in Figure 2.6. The details, however, can be a bit perplexing, and many practitioners state that there is not convincing evidence that justifies a shift of 1.5 sigma units versus, say, a shift of 1.3 or a 1.865 sigma. Six Sigma was begun internally at

Motorola. After the company won the Malcolm Baldrige Quality Award in 1986 and Six Sigma got out into the wider world, though, the principle of the shift got codified and became ingrained into the system. As a statistician, I think the overall concept is good, but I have a few quibbles with the doctrine of the 1.5 sigma shift in terms of statistical validity. Alas, as Black Belts, we inherit some terminology and must deal with it.

The 1.5 Sigma Shift

- Dr. Bill Smith at Motorola estimated that in the long run, a well behaved process can vary by as much as 1.5 standard deviations from its mean at any given time.

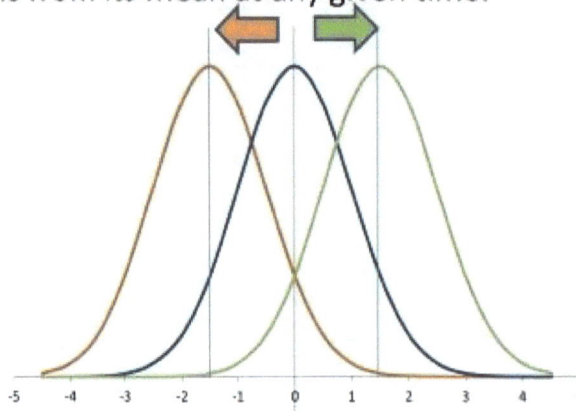

Arbitrary? Perhaps. But the 1.5 sigma shift has become the standard.

19

Figure 2.6 The 1.5 sigma shift

More specifically, from looking at Figure 2.6, we can see that the idea is that any well behaved, normally distributed process that is centered at the specification target value, can, over time, vary its mean. This variation can be as much as one and a half standard deviations from the original mean.

When we measure output from this process, we can be taking samples from the pristine process we think we have, or a process with a shifted mean. The worst-case scenario is taking samples from a process that is 1.5 standard deviations from the target. The takeaway is that even if we think we have a centered process, over time it is not necessarily stable in terms of its location. We should take this instability into account when we report defects. Hence, the defect counts under the 1.5 sigma shift reflect a possible extreme outcome for our process.

We are assuming our process output is normally distributed since defects per million opportunities (dpmo) calculations are based on area under the normal curve. We also note that the process is centered at specification target, which may or may not be the case. Last, we note that only the mean is shifting, and the standard deviation remains constant.

1.5 Sigma Shift, Illustrated

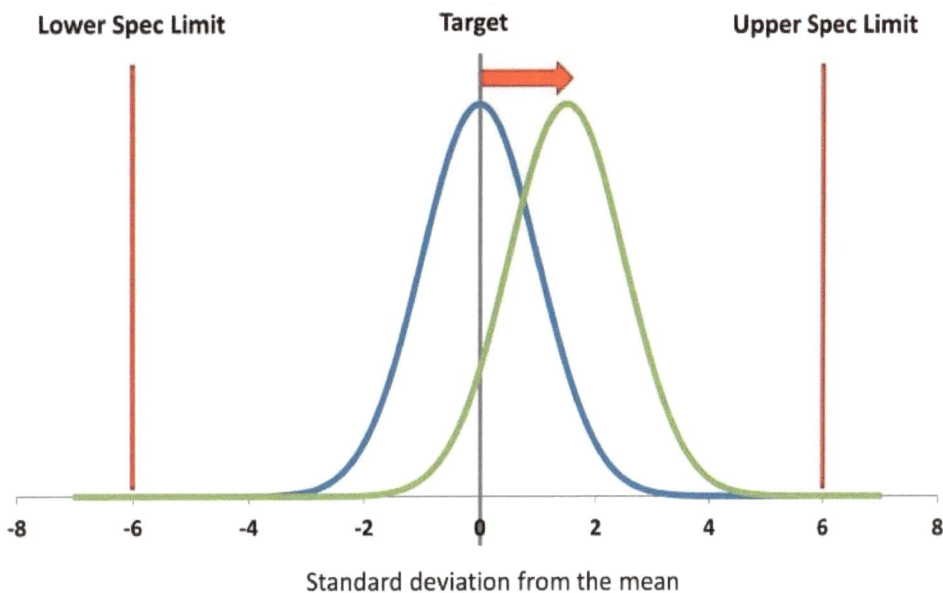

Figure 2.7 gives an illustration of the 1.5 sigma shift for a 6 Sigma level process. Note that the original process, in blue, is centered at target, and the specifications are located at +/-6 standard deviations from the target. This is a 6 Sigma centered process. Hardly anything falls out of specification: in fact, if we do the math for this blue distribution, we have only 2 parts per billion that are outside the specifications. If we apply the concept of the 1.5 sigma shift, we see that, under a worst-case, long-term scenario, on any day we take samples, we could in fact have the green process. (The process is

drawn shifting 1.5 standard deviations above the original mean, but it could have just as easily shifted by 1.5 standard deviations below the mean.)

With the shifted green process, we see that on the low end, there won't be many products falling out of specification since the shifted mean is now 7.5 standard deviations away from the lower

specification. But, on the high end, we will have some parts out of specification. For the green distribution, the upper specification maps to 4.5 standard deviations above the mean. This translates into 3.4 parts per million falling out of specification, or 3.4 defects per million opportunities. On the factory floor, we don't know if we have the blue process, the green process, or something in between when we take samples from the process.

To report short-term results, we do not apply the 1.5 sigma shift. We just have the blue process. In the long-term calculations, we do apply the shift. Because of the shift, long-term defect levels are always higher than short-term levels.

How does dpmo change with sigma? We can see in Figure 2.8 that there is not a linear relationship since the area under the normal curve is not linear with z. Note the y axis for dpmo uses a log scale.

DPMO by Long-term Sigma Level

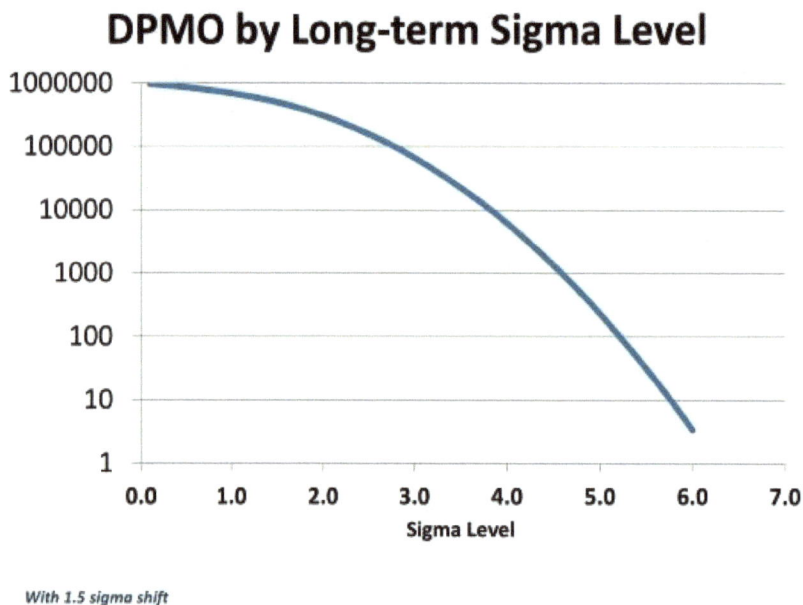

With 1.5 sigma shift

Figure 2.8 DPMO versus long-term sigma level

2.3 The Origin of Six Sigma

After Motorola won the Malcolm Baldrige Quality Award in 1988, word got out about its Six Sigma methodology, and other companies adopted the philosophy, refining it along the way to match their business needs. General Electric, Allied Signal and Honeywell all became Six Sigma companies. Most likely you know a Black Belt trained at General Electric – they are everywhere! In 2000 and beyond, more industries moved to Six Sigma, not just manufacturing, but service, banking, retail, and healthcare, among others. Despite the articles that are published each year reporting that Six Sigma is dead, it is still being adopted by companies, and is now combined with the Lean philosophy.

Six Sigma comes with its own specialized jargon. You might have wondered why we have the karate metaphor for quality positions. It seems a little hokey. We can thank Mikel Harry of Motorola for the martial arts references. As the story goes, a manager came up and told him that the Six Sigma tools were kicking the hell out of variation, and Harry imagined a karate kick in his mind. And, thus, the "belts" were born.

What do the various belts do? As summarized in Figure 2.9, Black Belts in large companies work full-time leading Six Sigma projects. Black Belts have knowledge of many analytical and statistical tools and are skilled at leading cross-functional teams. Many Black Belts are actively involved in creating and delivering quality training across the organization as well as mentoring Green Belts.

Green Belts often keep their regular jobs and work part-time on Six Sigma projects. As you know, Green Belts are knowledgeable about the DMAIC cycle, graphical methods, and Lean tools, and are capable of leading smaller scale projects. Most Six Sigma projects, especially in the beginning of implementation, will not require high powered statistical analysis.

Yellow Belts understand the Six Sigma philosophy and are familiar with some basic tools. They are poised to be active contributors to Six Sigma teams.

Finally, White Belts have an awareness of Six Sigma, its philosophy, and goals. These folks have been through, say, a half-day training session. Executives who must champion the Six Sigma effort are White Belt candidates. (Kubiak, 2009)

Six Sigma Roles: The Belts

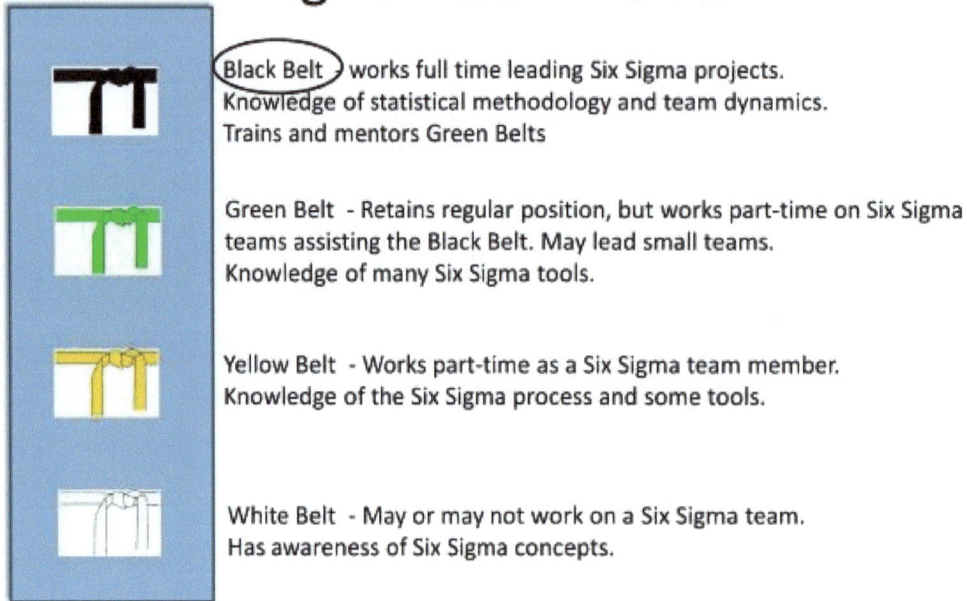

Black Belt works full time leading Six Sigma projects.
Knowledge of statistical methodology and team dynamics.
Trains and mentors Green Belts

Green Belt - Retains regular position, but works part-time on Six Sigma teams assisting the Black Belt. May lead small teams.
Knowledge of many Six Sigma tools.

Yellow Belt - Works part-time as a Six Sigma team member.
Knowledge of the Six Sigma process and some tools.

White Belt - May or may not work on a Six Sigma team.
Has awareness of Six Sigma concepts.

Source: Kubiak and Benbow,, The CSSBB Handbook, ASQ Press, 2009, pp. 17-19   25

Figure 2.9 Six Sigma roles

There are more Six Sigma roles. Master Black Belts must be very knowledgeable in statistical methods, team dynamics and project and program management. They are often members of the steering committee that chooses projects for the organization. Master Black Belts train Black Belts, and are considered the experts, the highest level Six Sigma resources in a company.

Each project team has a sponsor, or champion. This role is filled by an upper-level manager who can help remove any road blocks the team encounters. They are not active members of the team in terms of attending weekly meetings, but they do conduct the toll gate meetings and are in communication with the team leader.

A key role is that of the process owners. Process owners have the responsibility for a process, and the authority to make changes. It would be ideal if all process owners were at least Green Belts. Very often they are tasked to lead projects.

2.4 Six Sigma Features

Six Sigma is enterprise wide – it is not a program just for the factory floor. The front office, back office and janitorial staff should all be involved in Six Sigma efforts. Six Sigma is driven from the top down, with projects chosen by an executive-level steering committee. Decisions are data driven, and Six Sigma is oriented toward distinct, short-term projects. And there is an extremely high standard for quality that we set as our goal. We also must not forget that Six Sigma projects are specifically designed to bring more money to the bottom line, whether through increased market share or increased cost savings. The financials are an important part of the project charter and the validation of project success.

How did Motorola develop Six Sigma? Did they begin from nothing? Well, of course not. Quality existed well before 1986, going back as far as the medieval guild system. More recently, statisticians like Walter Shewhart, W. Edwards Deming and Joe Juran developed many data driven approaches to monitoring and improving quality. Motorola built upon this foundation and created a system that improved their quality and their profits. Is Six Sigma new, then?

One could argue that Six Sigma is not new because it uses existing tools, such as statistical analysis, statistical process control (SPC) charts, fishbone diagrams, process maps, and so on. It is team focused, but every other quality movement like Total Quality Management, and Quality Circles, were as well. Six Sigma is data driven, but that is certainly not new. And Six Sigma requires management involvement. Management always had to be involved for any quality implementation to take hold and thrive.

On the other hand, Six Sigma does have some unique features. The level of statistical sophistication required by Black Belts is unprecedented. Black Belts are expected to be able to perform hypothesis tests, multiple linear regression, designed experiments and capability analysis. Once relegated solely to the industrial statisticians, Six Sigma has now taken these techniques and democratized them, so that even Green Belts can explain hypothesis testing and alpha and beta errors.

The project approach is also a unique feature of Six Sigma. In contrast to continuous improvement, in which we incrementally improve a process and are never done, Six Sigma identifies a problem, and then a team goes in, uses DMAIC, saves money, and gets done. And then, the team disbands after celebrating, and the Black Belt moves on to the next project. It is more of a discrete process versus a continuum.

Finally, the integral emphasis on bottom line results sets Six Sigma apart. Other quality systems concentrated on reducing defects and improving customer satisfaction, as Six Sigma does, and it was assumed that the money would follow. Six Sigma specifically sets out to quantify exactly what the bottom-line lift will be for each project.

2.5 Lean and Six Sigma Together

Nowadays, we hear the terms Six Sigma and Lean used together. Lean originated in Japan and the Toyota Production System or TPS. Jeffrey Liker wrote an excellent book entitled *The Toyota Way* that explains the origins and concepts of this system (Liker, 2004). Six Sigma Black Belts have many Lean tools in their toolboxes. However, Lean is a quality philosophy unto itself. It is a common-sense approach that implements hands-on training instead of a lot of classroom time. Many tools require little to no mathematical knowledge with the result that practitioners are not studying statistics for weeks on end. Performing quick turnaround fixes, called Kaizen Blitzes, can offer quick dramatic results. Implementing a Lean approach is an easier sell than Six Sigma for these reasons. Many companies begin with Lean and then move into Six Sigma once a lot of the low hanging fruit has been harvested.

In contrast to Six Sigma, which emphasizes reducing variation, Lean concentrates on reducing waste. There are eight Lean wastes that follow the mnemonic TIM WOODS: Transportation, Inventory, Motion, Waiting, Overproduction, Over-processing, Defects and Skills. The key is to identify and eliminate these wastes to create more flexibility and to improve customer satisfaction. In addition, Lean subscribes to the concept of continuous improvement, in which we continue to incrementally improve processes.

Lean and Six Sigma can be combined, resulting in a synergy. By using the Six Sigma philosophy and Lean tools, organizations can reduce defects, waste, costs, and cycle times while increasing profits, quality, and customer satisfaction.

2.6 Six Sigma Benefits and Disadvantages

What can Six Sigma methods do for a company? Based on survey results, companies report that Black Belts save about $230,000 per project, and that the average Black Belt performs four to six projects per year.[1] That is quite a bit of savings. Now of course, depending on the size of your company, the result can vary.

General Electric, at the time a company included in the Dow Jones index, with thousands of Black Belts, estimated that they saved $10 billion dollars during the first 5 years of implementation. That payback certainly justifies the training hours for all those Black Belts.

Six Sigma does have its detractors, however. Some say that the Six Sigma methodology stifles creativity and impedes innovation. The DMAIC cycle is just a variation of any standard project management approach. But requiring excessive paperwork and justification for every change can become onerous. The company 3M impeded its R&D organization by imposing a Six Sigma framework on the inventors. Requiring too much paperwork and seeing everything through a Six Sigma lens is not the best approach. What appear to be failures of Six Sigma are really failures in implementation in which the Six Sigma system is applied too dogmatically.

Another criticism: statistics is hard! Certainly, most of the time in any Black Belt training course is spent learning statistical techniques. Six Sigma Black Belts must become mini-statisticians, and that is a challenging task. For the heavy lifting, an organization can hire a contract statistician to perform an analysis. However, everyone in the organization must have the mindset that decisions are best made through data analysis.

2.7 Causes of Six Sigma Failures

Many Six Sigma implementations do not perform as well as expected. In fact, many fail. Generally, Six Sigma failure is due to two things: leadership and implementation.

The executive leadership in an organization may treat Six Sigma as a fad, get excited about it, but then ultimately lose interest. Or the executives may mistake the Six Sigma methodology itself for a company goal. Implementing Six Sigma is not the goal, it is the means to achieve the company's goals concerning customer satisfaction, efficiency and bottom-line results. As a Black Belt, you will be able to raise red flags if you see these implementation pitfalls in your organization.

---

[1] See Mikel Harry and Richard Schroeder, (2000) Six Sigma, The Breakthrough Management Strategy Revolutionizing the World's Top Corporations

Implementation failures take the form of training too many people too soon, and not assigning them projects. On the other hand, a company can have too many employees working on Six Sigma projects, to the detriment of their other, customer-facing jobs.

Six Sigma projects may be chosen that are low-risk and hence low-reward. If a Black Belt is required to complete a set number of projects in a year, he or she may choose quick turnaround projects that really don't have much monetary or customer facing impact. The higher the reward, the riskier the project. Black Belts must be allowed to try these types of projects without fear.

There also might be too many projects going on at once, stretching resources too thin, and resulting in abandoned projects. In addition, projects may be selected that are not focused or are not tied to the strategic plan.

A common implementation failure is to assign root causes and solutions to projects prematurely. For example, a team could be given the project, "Reduce defects on production line #2 by installing a new welding robot." This project skips the first three DMAIC steps and lands on Improve. Assigning a root cause and solution without doing the D and M and A work may result in simply treating a symptom, or sub-optimizing the process. These suboptimal solutions are then incorrectly blamed on Six Sigma.

[In some cases, the solution to the problem is well understood. In this situation, a project would be classified as a JDI, or Just Do It, and there is no need to go through the DMAIC cycle.]

Finally, Six Sigma implementation can fail because excessive paperwork requirements can choke the system. Using streamlined templates and not making Six Sigma so bureaucratic will help with this. Six Sigma is a tool, and organizations can decide how best to implement it.

Chapter 3 Launching Projects

In this chapter we will review the methodology for choosing Six Sigma projects to deliver maximum impact. We will also cover the details for defining projects via the project charter.

Figure 3.1, first seen in Chapter 2, reminds us that Six Sigma projects are driven by the company's strategic goals which are directly related to meeting customer requirements and advancing the organization's business strategy. This diagram stresses that Six Sigma is a tool used to achieve company goals and is not an end unto itself. When first launching a Six Sigma program, there will be many candidate projects: a full pipeline of low hanging fruit which will easily fit into this circle.

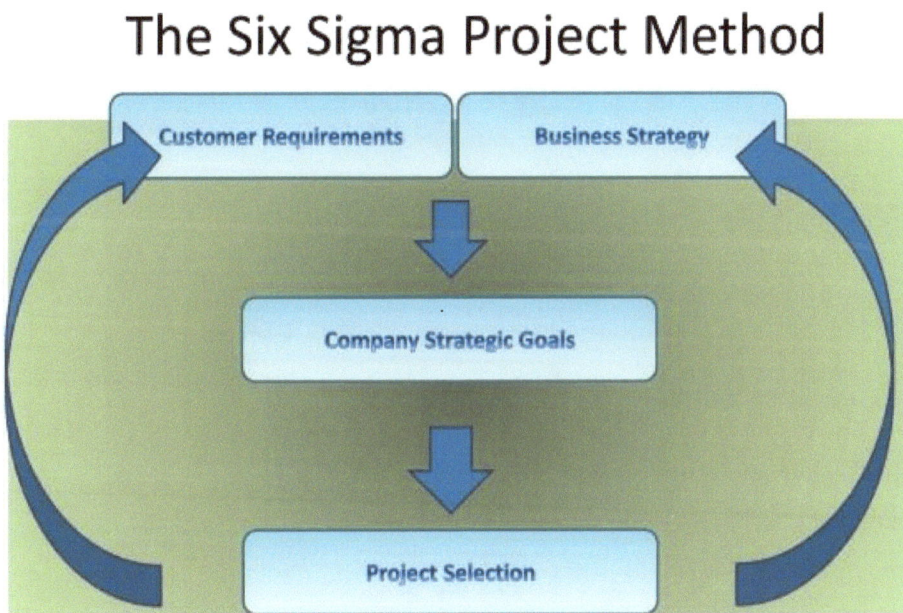

Figure 3.1 The Six Sigma project method

However, an organization cannot launch all the candidate projects at once. How does a company choose which projects it will perform and when? Many companies choose projects through a Six Sigma steering committee composed of executives and Black Belts or Master Black Belts. The committee will decide which projects to assign and when, based on several considerations.

As shown in Figure 3.2, the first consideration in choosing a project for implementation is to make sure that the proposed project aligns with the company goals. We want to make sure the project is not a Six-Sigma-for-Six-Sigma's-sake type of project. If a project does not address customer needs, then it is not a suitable candidate. Project duration is also a consideration — projects should be between two to nine months in duration. The committee should also choose projects that have a high cost-benefit ratio or return on investment (ROI). The complexity of the project is important: will a project require coordination across 3 divisions and multiple departments? If so, it might not be a viable candidate, especially for an initial launch project. Finally, does the organization have the resources to tackle the project? If it is expected that the project will require capital investment, or a full-time commitment of major equipment or engineering resources, is the company prepared for that?

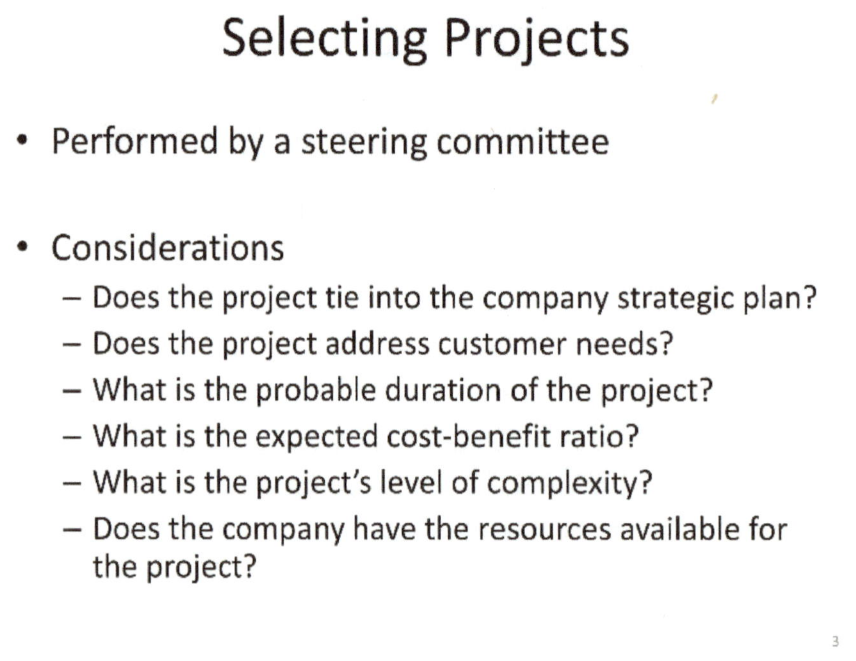

Selecting Projects

- Performed by a steering committee

- Considerations
    - Does the project tie into the company strategic plan?
    - Does the project address customer needs?
    - What is the probable duration of the project?
    - What is the expected cost-benefit ratio?
    - What is the project's level of complexity?
    - Does the company have the resources available for the project?

3

Figure 3.2 Considerations when selecting projects

When a company is launching its Six Sigma program, quick turnaround, high-return projects should be prioritized. Choosing these types of projects will accomplish several things. First, these projects will give the newly trained Black Belt, Green Belt or Yellow Belt the ability to test out his or her knowledge and put it to practical use. Second, these types of projects will set the team up for success, which will encourage them to tackle another project. Third, quick successes and results will provide proof of concept and will add to the momentum of the roll out. Luckily, there will be a lot of low hanging fruit projects to choose from when first starting out, so there will be many good options available.

3.1 Gathering Ideas for Projects

We can listen to the various voices of the organization to find pain points that can help us identify candidate projects. Along with the voice of the customer, the steering committee can tune in to these other sources of information:

- Voice of the Business: what is the business strategy, what threats and opportunities is the business currently experiencing?
- Voice of the Customer: what are the customer requirements, and what are customers willing to pay for?
- Voice of the Process: which processes are telling us that they are broken, or sub-optimized?
- Voice of the Employees
- Voice of the Environment
- Voice of Health and Safety
- Voice of Regulatory Bodies

3.2 Project Scope

As we've established, Six Sigma projects should take between two to nine months to complete. Long-term projects are often doomed to failure due to loss of interest or changes in the management team or company priorities. We need a project scope that is relatively small, not an unfocused project such as *Improve customer satisfaction*, or *Improve quality in the manufacturing plant*.

On the other hand, we don't want the project to be so narrow, so prescriptive, that the solution is either already known (a JDI), or it's incorrectly pre-identified. These projects are often phrased like *Improve process outcome X **by** doing Y*. Will the proposed solution address the root cause? It is difficult to say.

Finding the root cause of the problem is the purpose of the Six Sigma project, found after the project is defined, metrics identified, and data collected and analyzed. For example, we may be given a project phrased as *Decrease customer wait times **by** increasing staffing levels.* We could achieve shorter wait times by adding more staff, but is that just a Band-Aid solution? The cause of the slow service times might be an inefficient process, or poor training. We will not know until we dig into the problem with our Six Sigma methodology.

3.3 Project Prioritization

To choose among candidate projects, the steering committee can use a tool called a prioritization matrix. (Tague, 2005) Using the matrix, the committee rates potential projects on criteria it has agreed upon beforehand. These criteria are weighted according to their perceived importance. Using this tool gives the committee a way to choose projects based on data, not based on the loudest or most senior voice in the room.

The first step in creating a prioritization matrix is to get a consensus on the selection criteria. The criteria can be gleaned from the considerations shown in Figure 3.2, such as complexity of the project, cost benefit analysis, resources needed, etc. Once the criteria are decided upon, the committee then assigns weights to each. These two tasks can be performed in a separate meeting, before any proposed projects are considered.

To use the matrix, the committee goes through each criterion, and ranks each project, with a rank of 1 being the best. The project with the lowest total weighted score then becomes the top priority. There is a template in the online course companion spreadsheet that allows a committee to choose criteria, weight them using percentages, and rank projects against the criteria. The overall total scores are automatically calculated.

Figure 3.3 shows an example of how a priority matrix would work. Here, the committee has previously decided upon the criteria of speed, cost, and complexity. They felt that complexity was the most important consideration of the three and gave it a 50% weighting. Cost follows with 40%, and then speed at 10%. As a check, we can see that the weights add to 100%. The committee now must decide among 3 candidate projects, named A, B and C. Each criterion is considered separately. For Speed, the group believes that project C will be the "best," or fastest, to complete and assigns it a rank of 1.

Project Prioritization Matrix

Enter each criterion and its relative importance.

Enter candidate project names.

Rank the projects based on the criteria entered. Lowest number is best, or most desirable.

Criteria	Speed	Cost	Complexity		Total
Relative Importance (%)	10%	40%	50%		100%
Project Name					
A	2	3	2.5		2.65
B	3	1.5	1		1.4
C	1	1.5	2.5		1.95

Figure 3.3 Project prioritization matrix

Project A is next fastest and thus gets assigned a rank of 2, followed by B, which is given a rank of 3. For Cost, the group believes that projects B and C are tied for lowest cost. A tie is handled by taking the average of ranks 1 and 2 and returning a 1.5. This would be interpreted as lowest implementation cost. Project A is the highest cost and gets a rank of 3. As you are filling out the matrix, keep in mind that a rank of 1 is a positive thing – it is not the highest cost! For complexity, the group thinks that project B is the least complex, and hence it gets a 1 ranking. Projects A and C are tied for second place, so they each get a 2.5.

One quick check you can do to make sure you have handled ties correctly is to recognize that the ranks for each criterion must add to the sum of the n ranks. Here, we have n = 3 projects, so the sum of the ranks equals 1 + 2 + 3 = 6. That means that each column needs to add to 6. If there were four candidate projects, the columns would have to add to 1 + 2 + 3 + 4 = 10. There is a formula for quickly calculating the sum of consecutive numbers from 1 to n:

$$\frac{n(n + 1)}{2}$$

To add 1, 2 and 3, we note that n equals 3. In the numerator we have 3(3+1) = 12. We divide this result by 2 to find the solution of 6.

Based on the committee's rankings, we see that Project B earned the lowest overall score, thus it is the top priority based on the criteria and the weights. How did the committee approach the exercise of ranking the projects? With three projects and three criteria, it wasn't too onerous a task. But what if we had 8 criteria and 10 projects? The group might get bogged down on rankings, and the final decisions might be based on who had the most dominant personality. To avoid making this an arm-wrestling exercise, it might be helpful in that case to send out the matrix ahead of time and ask participants to rank the project independently. In the meeting, the group could then address only those cases in which ranks differed by more than two or three positions.

Of course, by ranking projects, we are using an ordinal measurement scale, in which the difference between ranks is not necessarily the same. Consider three runners competing in a 5K race. Imagine a very exciting finish, in which the second-place runner crosses the finish line just two seconds after the first. Twenty minutes later, the third-place runner finishes his race. Using the ranks 1, 2 and 3 does not give us an idea of the difference in finishing times of the runners, only their order. It is the same way with prioritization: we can't guarantee that the difference between 1st and 2nd is the same as the difference between 2nd and 3rd, and so on.

To avoid the limitations of the ordinal scale, the team may choose to assign ratings on a 1 to 10 scale instead of ranks in a prioritization matrix. Here it might make more sense to choose 10 to be the best, and 1 to be the worst rating. The team then rates each project based on this interval scale. This technique gives the team a bit more precision when distinguishing among projects. Ties are perfectly admissible and do not have to be treated in any particular way. The template in the course spreadsheet can accommodate either scoring method.

Note that the prioritization matrix is an example of an L-shaped matrix, in which two groups are compared. The first column of this matrix and the top row form an upside-down L, which is where the matrix type gets its name. For the prioritization example shown in Figure 3.6, the list of project names in the first column and the criteria and weights in the first two rows together form the upside-down L shape.

3.4 SIPOC Diagrams

Once we have the project assigned, we need to make sure our scope is in line with the expected project duration. We can do this by using a project SIPOC diagram. SIPOC is an acronym that stands for suppliers, inputs, process, outputs, and customers. The chart lets us quickly see what process steps, suppliers, inputs output and customers we will be considering in our project. Project scope can creep out of control, so a SIPOC is an effective way to remind the team what is in scope. The SIPOC provides a 30,000-foot view of the process: the flow chart does not need to be detailed. (Note that some quality professional call a SIPOC a COPIS diagram, because "the customer always comes first." I try to avoid these folks at quality conferences.)

To create the project SIPOC, identify the specific process boundaries. Where will the process focus begin and end? Denote this with a high-level explanation of the process steps that the team will address. Then, identify the outputs of the process, and the customers of these outputs. Go back and identify inputs and the suppliers of those inputs. All the items on the SIPOC will be deemed in scope for the project at hand.

The SIPOC can also be used outside the context of a Six Sigma project as a high-level process information sheet. In this application of the tool, the critical-to-quality characteristics of the inputs, process steps and outputs are identified.

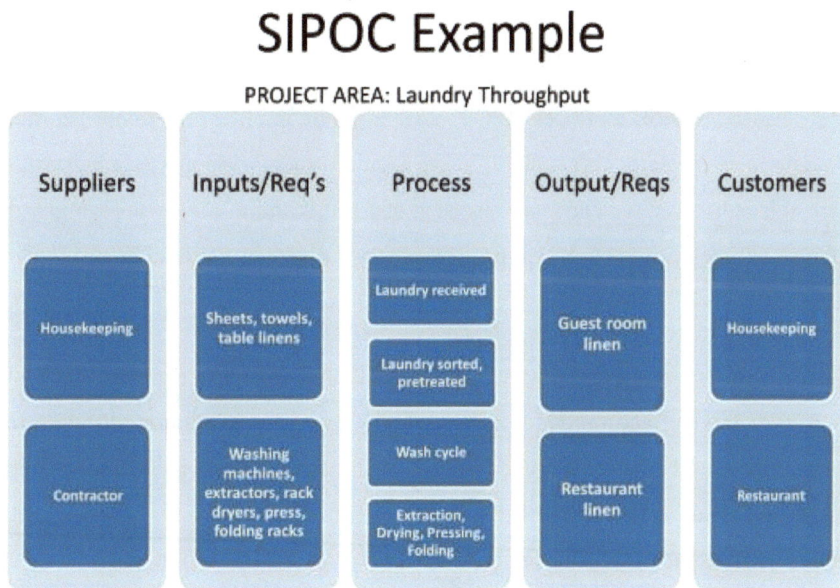

SIPOC Example

PROJECT AREA: Laundry Throughput

Suppliers	Inputs/Req's	Process	Output/Reqs	Customers
Housekeeping	Sheets, towels, table linens	Laundry received	Guest room linen	Housekeeping
		Laundry sorted, pretreated		
Contractor	Washing machines, extractors, rack dryers, press, folding racks	Wash cycle	Restaurant linen	Restaurant
		Extraction, Drying, Pressing, Folding		

12

Figure 3.4 SIPOC example

SIPOC Example

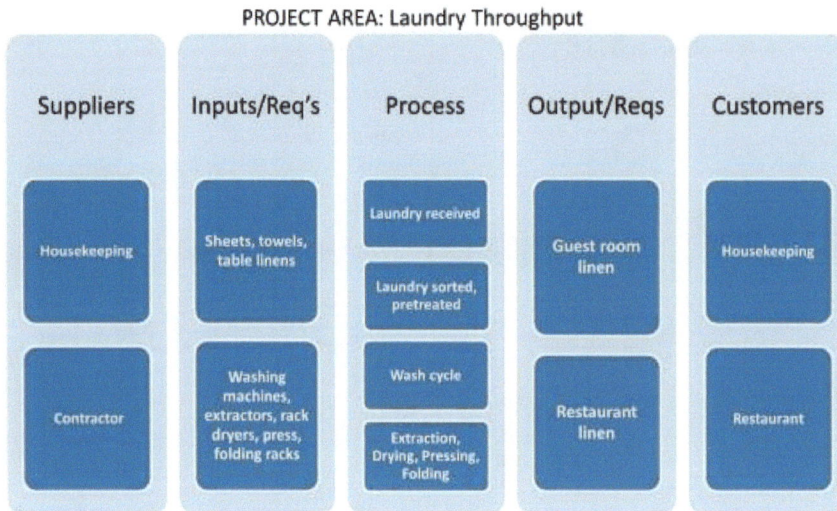

PROJECT AREA: Laundry Throughput

Suppliers	Inputs/Req's	Process	Output/Reqs	Customers
Housekeeping	Sheets, towels, table linens	Laundry received	Guest room linen	Housekeeping
		Laundry sorted, pretreated		
Contractor	Washing machines, extractors, rack dryers, press, folding racks	Wash cycle	Restaurant linen	Restaurant
		Extraction, Drying, Pressing, Folding		

12

Figure 3.4 shows an example of a project SIPOC chart based on the Mirasol Hotel case study found in Appendix A. The Mirasol Hotel is launching a Six Sigma project to increase laundry throughput at their commercial laundry facility. The team has been charged with improving the laundry throughput. What boundaries will they put around their project so that it does not increase in size and become too big to handle?

Where will the process begin and end for the purposes of the project? The team will start with laundry being received, then the sort and pretreat steps, wash cycle and the extraction, drying, pressing, and folding stations. The outputs the project will concentrate on are the guest room and restaurant linens. Customers of these outputs are housekeeping and the restaurant, respectively. Looping back, what are the inputs to the process steps that are in scope for this project? The inputs are sheets, towels, and table linens, supplied by housekeeping. Note that this supplier is also a customer.

What the SIPOC Did Not Include

PROJECT AREA: Laundry Throughput

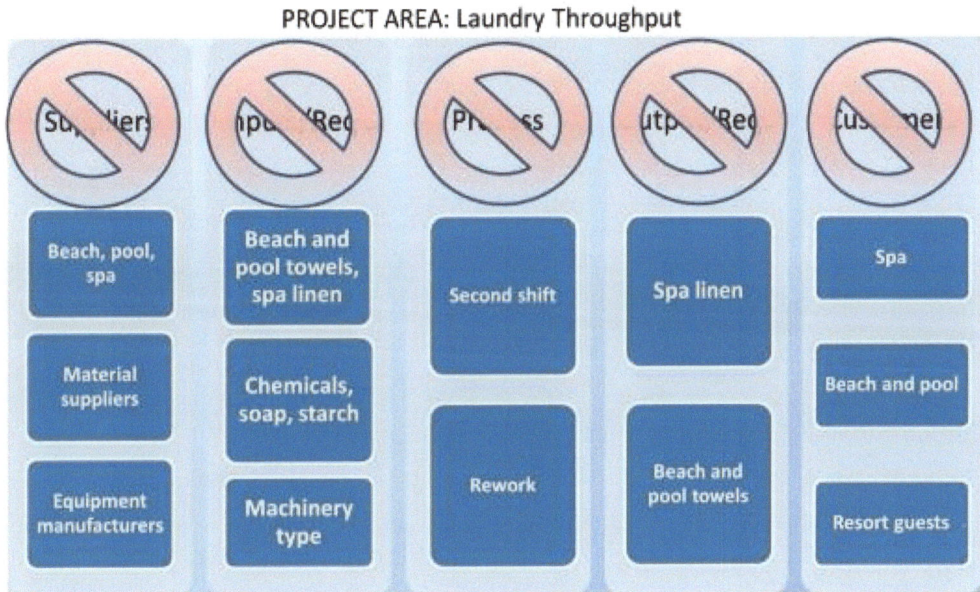

⊘ Supplier	⊘ Input/Req	⊘ Process	⊘ Output/Req	⊘ Customer
Beach, pool, spa	Beach and pool towels, spa linen	Second shift	Spa linen	Spa
Material suppliers	Chemicals, soap, starch			Beach and pool
Equipment manufacturers	Machinery type	Rework	Beach and pool towels	Resort guests

13

Figure 3.5 Anti-SIPOC chart

Another input is the leased laundry equipment: the washing machines, extractors, dryers, presses, and folding racks. These are supplied by the laundry contracting firm.

The resulting SIPOC is a one-page document that lists everything that is in scope for the project. It is valuable during the project to make sure the team does not veer off into another direction or add extra focal points to the project.

SIPOC seems like a great tool, but I have always thought it was a bit incomplete. For me, it is more helpful to have a map of what is in scope, the SIPOC, coupled by a map of what is *not* in scope, explicitly stated. I'll call this companion the Anti-SIPOC for want of a better term (Figure 3.5). This Anti-SIPOC states what is NOT in scope for a project. For example, the team will not focus its attention on the second shift operations in the laundry facility, nor will it include the rework process at this time. The project will not concentrate on the throughput for beach, pool or spa linen, or the spa, beach and pool or resort guests as customers. The project will not investigate the soap, starch or chemicals used in the process, and no new equipment is in the budget. Accordingly, the equipment manufacturers and material suppliers are not in scope as suppliers. I think the Anti-SIPOC, used in conjunction with the SIPOC, will provide a more complete picture of the project scope and help teams stay on track. There are templates for both diagrams in the spreadsheet.

3.5 Crafting Problem Statements

For each Six Sigma project, a project charter is created. This is the foundational document that explains what the project is about, what the scope of the project is, the basic timeline, project goals, team members, and a projection of money saved, among other things. Having a project charter helps the team focus on the project goals.

One of the principal elements of the project charter is the problem statement. Ideally, this should be one sentence. If you can state the problem in one sentence, you have really thought it through. Crafting the problem statement is a little like writing by numbers. The statement needs to have these elements (Wortman, 2004):

- How long has the problem existed?
- What is the measurable item? If we cannot measure it, how will we collect data, and how will we know if we have improved?
- What is the performance gap - where are we now and where should we be? This means we need to know the specification, or standard of quality for the metric.
- And what is the business impact? This is the "so what" element. Why is it important to be doing this project?

The following problem statement has all four of these elements: *Over the past six months, the number of surface defects on door panels has increased from 5/1000 to 60/1000, resulting in increased rework and warranty claims.* This is not necessarily good problem to have, but certainly a well-crafted problem statement.

The statement shows that the organization has clarity about this problem. Note that there is data in the statement, and this data has been tracked historically. Note too that the team might need to collect process data to be able to fully define the problem.

Writing a problem statement is harder than it looks. I have seen problem paragraphs! Make sure the causes or proposed solutions are not included in the problem statement. Finding the root cause and solving the problem is what the Six Sigma project is for. If the solution is obvious, then that is not a Six Sigma project, it is a Just Do It project. Make sure the item is measurable, and put in data, not conjecture.

Evaluate the following problem statement:

Customer complaints have always been higher than they should be, resulting in lost sales.

Recall that we are looking for four specific elements: how long has the problem existed, what is the measurable item, what is the performance gap, and what is the business impact.

- Customer complaints – that is the measurable item.
- Have *always* been higher than they should be? Are they saying the problem has existed since the beginning of time? We might want to go look at some data to see when the problem has started.
- …than they should be. Sounds like there is no clear requirement. It will be hard to know if we are successful in the project if there are no clear specifications put on the measurable item.
- …resulting in lost sales. That is a business impact. If we could put a dollar figure on that, it would be even better. This is the project rationale.

We have established that there are some problems with the first statement.

Let's evaluate another statement:

Reduce sealing defects from 5% to 1% by replacing the sealing machine with an improved model.

This sounds like an action statement, not a problem statement. We have a goal of reducing the defects from 5% to 1%, and we also have a solution in the statement: replace the sealing machine. Now, if that is what the team is going to do, then they should go ahead and replace the sealing machine. But this is not a Six Sigma project. However, it might be that a solution to the problem has found its way into the problem statement and buying the sealing machine would not necessarily be the best option. The team might be able to refurbish its existing machine or change the settings to achieve defect reduction. To sum up, this statement is not a problem statement: it has a goal in it, and it also includes a solution to the problem.

Now for the last example:

Over the past year, hotel bookings have decreased by 10% due to rude employees, resulting in a $400,000 budget shortfall.

- Over the past year… The "how long."
- …hotel bookings… That's the measurable item.
- …have decreased by 10%... That is the performance gap.
- …resulting in a $400,000 budget shortfall… There is the business impact. Sounds great.

But there is an added bit of information:

- …due to rude employees …

Whoops! Here the problem is being attributed to a cause without any analysis. How do we know the reduced bookings are due to rude employees? Isn't this pointing fingers? What evidence is there? Why are employees rude? Are they overworked? Not trained properly? Not paid enough?

Dr. Deming stated that 94% of all problems, defective goods or services come from the *system*, not from a careless worker or a defective machine. In other words, it is the system that must be changed, not necessarily the machines or employees. The system is of course controlled by management.

Make sure that problem statements do not assign blame. Dig deeper to get to the systemic root cause. If we strike the statement about rude employees, this last statement is a well-crafted problem statement.

Let's review some more problem statements:

Over the past six months, customer dining complaints have increased by 20%, resulting in the loss of at least 10 loyal customers.

- How long – the past two years
- Measurable item – customer dining complaints
- Performance gap – increased by 20%
- Business impact – resulting in the loss of at least 10 loyal customers,

This looks like a good problem statement. Not a good problem to have, but a well-crafted problem statement.

Try this one:

Order fulfillment time for product X has increased from 3 to 12 weeks in the past year, resulting in $100,000 in expedited shipping fees.

- Measurable item – order fulfillment time for product X
- Performance gap – increased from 3 to 12 weeks
- How long – in the past year
- Business impact – $100,000 in expedited shipping fees

This is another well-crafted problem statement.

3.6 Goal Statements

The goal statement goes hand in hand with the problem statement. The goal statement should be a SMART goal: specific, measurable, achievable, realistic (or relevant), and time bound. This is not the time for stretch goals. We want the Six Sigma goal statement to be something the team can realistically accomplish. It should also be time bound, which comes naturally here, since the Six Sigma project will be anywhere from two to nine months in duration.

For example, for the door defect project, a goal statement could be:

Reduce surface defects on doors from 60/1000 to 5/1000 within 6 months.

The statement is specific and measurable. It is achievable and realistic (or relevant) since we were running at 5/1000 previously. And it is time bound.

We need to be careful in choosing a goal statement – we do not want to sub-optimize our process. For example, we don't want to reduce defects on the doors on the one hand but increase rework on the line. We will explore this more in Chapter 5, Choosing Metrics.

3.7 The Project Charter

In the course spreadsheet, there is a project charter template in which the problem statement and goal statement are entered, along with the desired business results, or dollars saved. (Figure 3.10) There is space provided to describe the project scope, but the team leader can simply attach the SIPOC and

Anti-SIPOC diagrams to the charter. Team members and estimated resources needed are listed. A high-level schedule of when each of the five DMAIC phases will be complete is also included. The charter is signed by the project champion, the process owner, and the finance department, who vets the potential saving of the project. This will become the team's guiding document throughout the project cycle.

SIR Project Charter

Project No.		Black Belt / Green Belt / Yellow Belt:	
Project Title:			
Project Leader:		Telephone Number	
Project Champion:		Telephone Number	
Start Date:		Target Completion Date	

	Element	Description			
1	Business Process:	Describe the business process in which opportunity exists.			
2	Problem Statement:	How long has the problem existed? What measurable item is affected? What is the performance gap? What is the business impact?			
3	Project Goal:	Describe the project goal. (SMART)			
4	Objective:	What improvement is targeted and which is/are the metric(s) to be used to evaluate progress?	Metric	Current State (Before)	Proposed Future State (After)
5	Business Results ($):	Include a breakdown of projected cost savings. (Attach a separate sheet if necessary.)			
6	Team members:	List the core team members and any SMEs who may be brought in to help the team.			
7	Project Scope:	Which part of the process will be investigated? What are the project boundaries? Attach a SIPOC diagram.			
8	Resources needed:	List resources that may be needed to accomplish project goals.			
9	Schedule:		Preliminary Plan	Target	Actual
			Start Date		
			Define		
			Measure		
			Analyze		
			Improve		
			Control		
			Completion Date		
10	Approvals		Signature		Date
	Process Owner:				
	Project Champion:				
	Finance:				

Figure 3.6 Project charter

Chapter 4 Capturing the Voice of the Customer

To increase customer satisfaction, we must first identify our customers, and determine what they want. Who is our customer? The list is varied and most likely long. A customer can be the end user of our product or service, certainly, but there can be many customers in between: intra-department, inter-department, distributors, retail outlets, and so on. Organizations need to identify these customers to assure that their expectations are met.

4.1 Sources of Customer Data

Once we identify our customers, the next step is to determine what each of our customers requires. This is called collecting voice of the customer data which helps us know what our customers care about and what they are willing to pay for. Companies can then align our corporate goals with these customer needs.

How do we go about collecting voice of the customer (VOC) data? If we collect VOC from only one source, or one customer segment, we might get a skewed vision of the truth, as the late British comedian Benny Hill aptly pointed out: "Just because no one complains doesn't mean that all the parachutes are perfect."

We can analyze warranty data to find out how our products are failing or tally the number of returned items to determine what products do not meet customer expectations. These are inexpensive, indirect ways to gather information from the customer since organization already track it. However, using this approach, we will only glean information on defects, and not what pleases the customer. We can also look at complaint cards or feedback cards. These require the customer to respond to us. Again, we will hear from customers who are very unhappy or extremely pleased, which is useful information, but the majority of customers who are somewhere in the middle will not be represented by this technique.

Focus groups allow companies to bring in current customers or a customer target group using an incentive. Typically, participants in the focus group will interact with the company's products and give feedback on what they liked and did not like. Focus groups can be expensive to run, and the information may or may not be very useful. As my friend who works in marketing at a top auto maker tells me, people can home in on one issue and ignore the rest. In her example, participants might not offer much actionable feedback on a new engine and drive train, but they can go into exhaustive detail when discussing the car's cup holders.

Another source of customer information is the survey. Surveys are ubiquitous. With online survey software, creating and distributing surveys is easy. However, designing high quality, actionable surveys takes quite a bit of planning and knowledge.

Before creating the survey, the designer must identify the population of interest and the purpose of the survey. For example, a population of interest could be current customers in the 25-35 age range, and the purpose could be to gather information on which additional services those customers desire, or how satisfied they are with the current product offerings.

Once the responses are collected, the designer must consider how well those responses represent the population of interest. Who is filling out the surveys? Are respondents at the extremes of the distribution (meaning very unhappy, or very happy), leaving out the majority middle? How valid are the responses? Are customers tending to choose the middle response all the way through the survey simply to get it done? Our do they submit incomplete surveys? Good survey design and piloting can help mitigate these types of response problems. Survey design is covered in more depth in section 4.4.

Tracking trends on social media sites such as Twitter, Snapchat and Facebook can give a company insight into what customers are saying about their products and services. It is also an effective way for a company to communicate with its customers about specials, new products, and events.

Another means of collecting customer data is point-of-use observation, in which customers are tracked as they interact with the product or service. Examples include usability testing for software, in which a group of people that represent the customer base are asked to perform a set of tasks, and their clicks and stumbles are recorded to try to improve the usability of the interface. Websites track customer behavior and record the source of traffic, number of pages visited, time on page, and so on. In much the same way, customers are observed in supermarkets and tracked on what aisles they go to and in what order, how long they spend in each aisle, and on whether they linger at a certain spot, say, the cereal section. Length of the checkout lines are tracked by day of week and time of day to help with scheduling. In a call center, average wait time, average rings to answer, and average call times are constantly being tracked.

4.2 Kano Analysis

To help us understand customer requirements, we can use the Kano Model to classify each feature in a product or service as either a satisfier, a must-have, or a delighter.

A *satisfier* is a feature in a product or service that is known to the customer and can be articulated by the customer. The feedback from surveys, complaint cards and focus groups often revolves around satisfier features. Satisfiers are also referred to as *one-dimensional features* or *performance features* in which having more is better. Examples include computer speed (faster the better), the number of sports channels in a cable package, cell phone memory, discount percentages on clothing, or hours of operation for a drugstore.

A *must-have feature* is expected and known to the customer. A must-have is a basic requirement of a product or service. Because these features are so basic, they are not voiced by the customer. Accordingly, these are features that would not be discussed in a focus group situation. It is said that a must-have feature is only noticeable in its absence: if it is present, the customer won't provide feedback, but if it is absent, it will generate a complaint. For example, when walking into a hotel room after a long day on a business trip, you expect to see a bed and a separate bathroom. A hotel having these room features will not make you happy, it is what you expect. If, however, your hotel room had a hammock and a shared bathroom down the hall, it is very likely that you would check out and find another hotel.

Finally, a *delighter* is a feature that customers are unaware of and therefore not voiced by them. Think of these as "Oh wow!" features. Delighters will not usually be identified in focus groups. Instead, the innovation created by the marketing and product development departments will create features that will delight customers.

Henry Ford made a case against using focus groups to develop innovative products, as expressed in Figure 4.1. Steve Jobs was also not a big proponent of using focus groups to harvest ideas for new products. He said it was the job of his engineers and designers to come up with the next new things that nobody had thought of. Customers would not do that, his employees would.

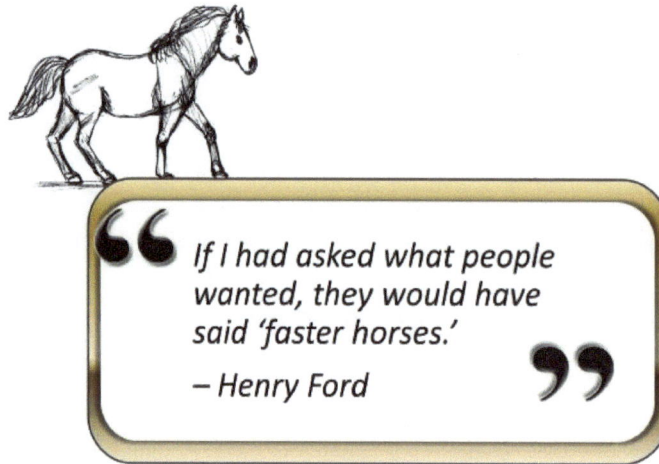

Figure 4.1 Henry Ford quote

Figure 4.2 shows a perceptual map showing the interplay among the three categories of product or service features. The blue line represents what happens with one-dimensional (satisfier) features. When the feature is not present, or nor performing well, customer satisfaction is low. As we increase quality, or increase the presence of a feature, the customer satisfaction grows in a linear manner: the more the better.

Perceptual Maps: The Kano Model

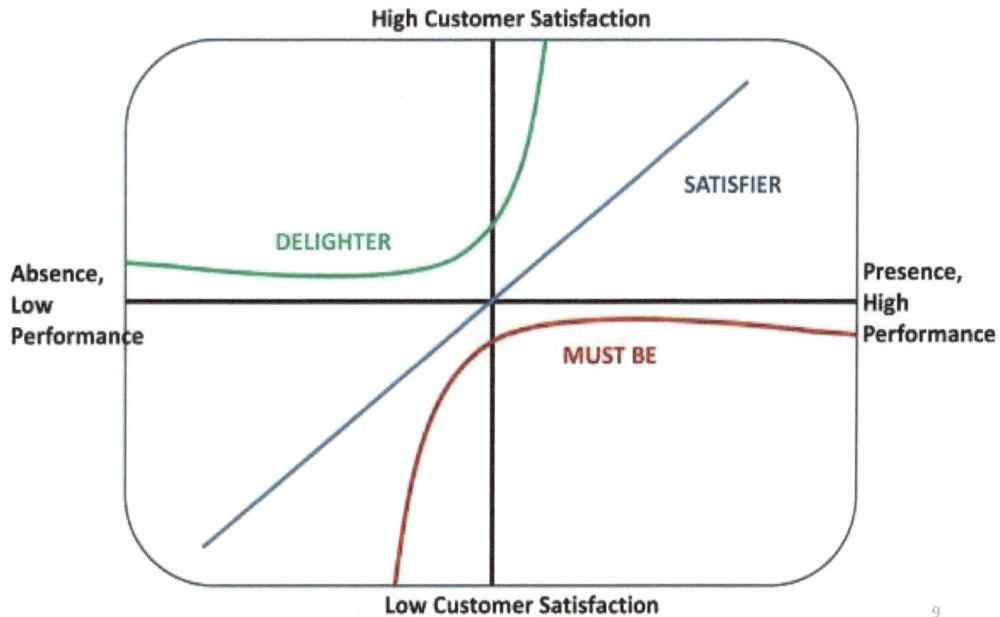

High Customer Satisfaction

SATISFIER

DELIGHTER

Absence, Low Performance

Presence, High Performance

MUST BE

Low Customer Satisfaction

9

Figure 4.2 Perceptual map of Kano model

For must-have features, if the feature is not there – if there are no chairs at work, or there is no Wi-Fi in the hotel, for example – customers are very unhappy. As soon as the feature is introduced, satisfaction improves, but stays at a neutral spot. A company will never get a pat on the back when it supplies a must-have; it can only get a complaint when the feature is not there.

Delighters work in the opposite way. When a delighter is not present, there is no influence on satisfaction since the customer was not expecting it anyway. As soon as the delighter is introduced, though, customer satisfaction increases rapidly. Examples can be technology oriented such as a self-parking feature of a car, or service oriented like same-day, free delivery or a complimentary glass of champagne at check-in at an upscale hotel.

Of course, the adage is true: a luxury once tasted become a necessity. Recognize that a delighter has a shelf life, and after a while, the feature might become a must-have once customers start expecting it.

4.3 Translating VOC to CTQs

Once we collect VOC data, we must translate it into something we can use in the business. Our customers do not give feedback in terms of our process variables or corporate goals. In fact, the feedback we receive is often vague and not necessarily actionable. We must take these vague statements and translate then into Critical-to-Quality requirements (CTQs). These CTQs state what the customer expects in terms of quality, delivery, cost, performance, and so on. Next, we will identify key process output variables (KPOVs) that measure the CTQ. Then, we will specify what range the KPOVs should fall into by using functional requirements, or specifications. Finally, we can determine the key process input variables (KPIVs) that link to the outputs. All this jargon sounds like alphabet soup right now. Let's start with studying the CTQ Mapping diagram in Figure 4.3 to make sense of this process.

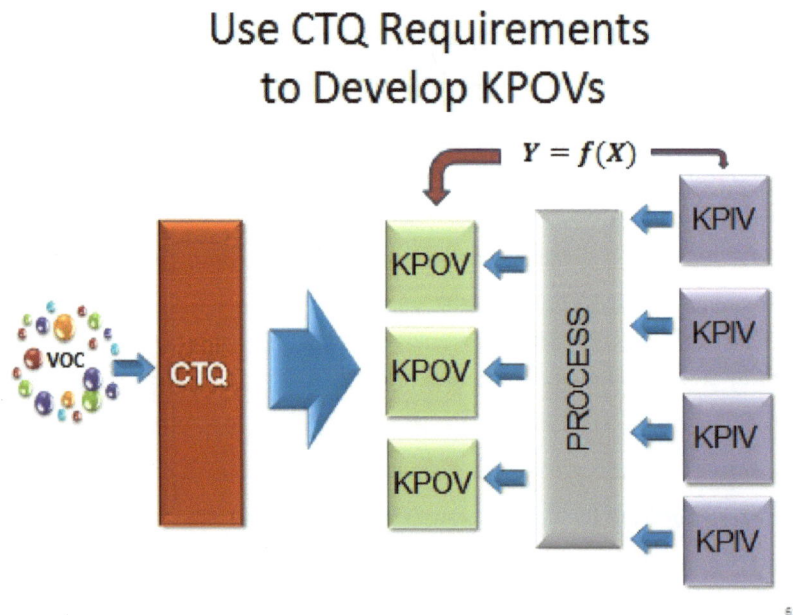

Figure 4.3 CTQ mapping

Once we get the CTQs from the VOC feedback, we can name key process output variables that link to the CTQ. If the KPOV has a value within the specified functional requirements, we know the CTQ will be met, which in turn will satisfy the customer. It makes a chain. Later, we will learn to go farther down the process. If we can identify key process input variables that will give us the KPOV values we want, this is the best since controlling inputs will reduce scrap and rework. We will know what input values will create a product that has the correct KPOV values, which will satisfy CTQs and the customer. This is called determining the relationship $Y = f(X)$, (Y equals f of X) in which Y is the output and X is the input or inputs.

We can illustrate this process using an example of a sandwich shop. Customers who call in orders for pick up or delivery were targeted for a VOC study. The shop collected complaint cards and comments from surveys over a three-month period. Figure 4.4 shows six statements that are representative of the most frequent types of comments.

We will take this customer feedback and turn it into something the sandwich shop can use to improve customer satisfaction. To determine the CTQs, we can take each comment and identify the overarching principle it touches on. We can be helped by using the prompt *"Our customers deserve,"* or, *"Our customers require."*

For example, the first feedback statement is that ordering is too complicated. We can turn this into a more general critical-to-quality requirement of *"Our customers require a streamlined, organized menu."*

CTQs

Customer Feedback	CTQ
Ordering too complicated	Streamlined, organized menu
I didn't know there was a lunch special when I called in	Information on specials at time of order
Lettuce not crisp	Fresh, quality ingredients
You gave me the wrong bread three times! Rye is not the same as wheat, people.	Error free orders
The pick up time you give is never right. I either have to wait, or my stuff is cold.	Accurate pick-up times
Delivery guy is unfriendly	Friendly customer service

Our customers deserve...

Our customers require...

15

Figure 4.4 Sandwich shop CTQs

"I didn't know about the lunch special" turns into "*Our customers deserve information on specials at time of order.*"

The lettuce not being crisp, or any type of comment about the food can turn into, "*Our customers deserve fresh, quality ingredients.*" We are taking specific feedback and turning it into general customer requirements.

"You gave me the wrong bread" becomes, "*Our customers deserve error free orders.*"

For the next comment about pick up times, we can translate it into "Our customers deserve *accurate pick-up times.*"

Finally, the comment about friendliness can translate into, "*Our customers deserve friendly customer service*" as another CTQ. Now we are getting somewhere.

Continuing along the chain shown in Figure 4.3, we can link key process output variables to the CTQs. In other words, what metrics should we track to make sure the CTQ is being satisfied? Note that we are looking at output variables right now. These are metrics we use at the end of the process, once the product or service is complete. From there, we can develop a functional requirement, or a specification, on the KPOV. In other words, what range of values for the metric will assure that the CTQ is satisfied?

For the CTQ of a streamlined, organized menu, we can measure the time it takes for the customer to order his or her sandwich. (Figure 4.5) The thinking here is that fast order time is related to ease of ordering. What should the order time be? This shop set the time to be under 30 seconds. If the order times for a sandwich are longer than that, the shop knows the menu and ordering process is too complicated.

CTQ	KPOV	Functional Requirements
Streamlined, organized menu	Customer time to place order	Less than 30 seconds for single sandwich orders
Information on specials at time of order	Number of lunch specials ordered by phone vs. number of lunch specials ordered in person	Difference less than 10%
Fresh, quality ingredients	Food complaints	Complaints < 1 per month
Error free orders	Order errors	Order errors < 1%
Accurate pick-up times	Time between stated pick-up and order ready	Difference within 5 minutes
Friendly customer service	Customer service complaints	Complaints < 1 per month

16

Figure 4.5 Sandwich shop functional requirements

"Information on specials at the time of order" is the next CTQ. How will we know we are meeting that requirement? We can create a script for the order-takers, so they don't forget to mention the daily specials. Then, we can track the sales of the specials for in-person versus telephone orders. If it is about the same, we know that the call-taker informed the customer of the special. To be more specific, if the difference is less than 10%, we can assume that callers knew about the special.

For the fresh ingredients, the shop will track the number of complaints about the food. The functional requirement is to receive fewer than one complaint per month.

"Error free orders" is also straightforward. They will track order errors and want the frequency to be less than one percent.

For "accurate pick-up times given to callers," the shop will track the time given over the phone versus the actual time the order was ready. They want the difference to be five minutes or less.

Friendly customer service will be tracked via customer service complaints, with the requirement of fewer than one complaint per month. Note that the KPOVs measure outcomes after the fact: after the order is taken, or the sandwich is made or delivered. These are not proactive metrics, but they are a useful way to see if the CTQs and hence the customer requirements are being met.

As shown in Figure 4.6, we can step farther up the chain and associate the KPOVS we have identified with specific processes in the shop. For example, the KPOV of "customer time to order" is associated with the menu design process and the order entry process. The KPOV is an output of these processes. For order errors, the order taking, order entry, sandwich prep and packaging processes are all involved. In other words, if one of the KPOVs were outside the allowable range, which process, or processes, would be investigated? If the shop gets many customer service complaints, it will concentrate on the order taking, counter and delivery processes.

KPOV	Process	KPIV
Customer time to place order	Menu design and order entry	$y = f(x)$ $KPOV = f(KPIV)$
Number of lunch specials ordered by phone vs. number of lunch specials ordered in person	Order taking	
Food complaints	Supply chain and storage	
Order errors	Order taking, order entry, sandwich prep, packaging	
Time between stated pick-up and order ready	Order entry, sandwich prep, packaging	
Customer service complaints	Order taking, counter, delivery	

Figure 4.6 Sandwich shop process identification

There also is a column for Key Process Input Variables (KPIVs). These are the inputs into the processes identified. The purpose of the Six Sigma project would be to identify which inputs are key, and what their levels should be. If we can control inputs, then we can assure the proper

outputs, and satisfy CTQs. We would like to move our focus to as early in the chain as possible. Controlling KPOVs is fine, but the labor and materials have already been used, and the customer has already been made unhappy. If we can control inputs, however, then we can better assure good outcomes for the customer. This link between the inputs and the outputs is called finding Y = f (X). This is the purpose of the Analyze phase, where we identify root causes of the problems and key input variables.

4.4 Developing Surveys

To gather customer data, a company may decide to conduct a survey. As pointed out earlier in section 4.1, it is very easy to develop a survey using Survey Monkey, Vendavo, Qualtrics, or any number of survey software providers. It is not as easy to create a well written, unbiased survey that will yield actionable results.

The first step in survey design is to determine the purpose of the survey. Who is the target audience, and what is it that the company wants to know? How can it assure that the knowledge gleaned is actionable? How will data be collected: through an email survey, a paper survey, or a survey conducted by an interview? Time spent addressing these design issues will result in a cost-effective survey that yields actionable results.

The target audience and purpose of the survey will then determine the design. Keep the following design tips in mind when creating a survey. First, start with the meat, and leave the demographic questions until the end. People have short attention spans, and they are busy. Do not waste time asking for a person's gender, age, and income right from the start. Instead, present the most important questions first. For example, consider Mirasol Hotel from the case study in Appendix A. The hotel might want to know what activities guests would prefer, or how guests rate their rooms on comfort.

Many surveys ask demographic questions, but will those demographics be used? If a company does not plan on cross tabbing the survey results by demographic category, then those types of questions just take up valuable real estate on the survey form.

Keep the survey short. We have all been sucked into surveys that seemed to have no end. Providing a percent completion scale on the online survey helps people pace themselves.

Pilot test the survey. Ask another department to take the survey and give feedback before sending it out to the customer. You will be surprised at the things you didn't catch the first time. Once the results are in, publish them on a web site. If the customer took time and energy to fill out the survey, they deserve to see the results. Transparency is best.

Designing good survey questions takes time and attention to detail. There are many types of mistakes an analyst can make, such as creating leading questions, loaded questions, double-barreled questions, overlapping response choices, non-exhaustive choices, and non-specific questions.

A *leading question* does just that: leads a respondent to a choice. This type of question is sometimes used on purpose in political surveys when the organization wants to show a high percentage of respondents agree with an issue. But, if you really want to know what your customer or employees think, avoid leading questions. For example, if the Mirasol Hotel asked its employees:

Leading Question: Lean Six Sigma implementation can save our hotel millions of dollars and allow it to grow. Do you think that employees should be trained in Six Sigma techniques?

The first sentence sets respondents up to of course answer affirmatively to the question. It's better to take out the back story and just ask the question:

Better: Should our employees be trained in Six Sigma techniques?

A related type of poorly worded question is the *loaded question*. Here, an assumption is made in the question and the respondent is forced to answer. For example, asking when a resort guest is planning on taking a dolphin cruise and then listing times assumes that the guest is in fact planning on taking the cruise. Respondents are forced to choose a preferred time when they might not ever sign up. This approach will give a false sense of demand for the cruises if the purpose is to create a schedule.

Loaded question: What time is your first choice for a dolphin cruise?

- o   Morning [10 am]
- o   Afternoon [3pm]
- o   Dinner [6pm]
- o   Sunset [8pm]

Better: What time would you choose for a dolphin cruise?

- o   Morning [10 am]
- o   Afternoon [3pm]
- o   Dinner [6pm]
- o   Sunset [8pm]
- o   Not interested in a cruise

A *double-barreled question* asks for two opinions at once. For example:

Double-barreled question: Do you think Six Sigma is easy to implement and worthwhile?

A respondent might think Six Sigma is worthwhile, but not very easy to implement. How does he or she respond? The data from this question will not be useful. It is better to break the question into two separate pieces:

Better: 1. Has Six Sigma been easy to implement?

2. Has the Six Sigma implementation been worthwhile?

A frustrating problem for survey takers is answering questions with *overlapping responses*. The categories are not mutually exclusive. Again, this will result in the data being suspect. For example:

Overlapping responses: What is your income?

- $25,000 or less
- $25,000 – $50,000
- $50,000 – $75,000
- $75,000 or more

Which category does a respondent choose if she makes $50,000 per year? The better choice is to assure that each category is mutually exclusive.

At the other end of the spectrum, we have the *non-exhaustive choices*, where a respondent's answer is not included in the listing. For example:

Non-exhaustive choices: What is your house worth in today's market?

- **$200,000 – $249,000**
- **$250,000 - $499,000**
- **$500,000 – $749,000**
- **$750,000 or more**

If a respondent's house is worth $185,000, how does he respond to this question? He might think that the survey was not meant for him and stop altogether.

Finally, we have *non-specific questions*. The question will yield data, but it will not be trustworthy since the question is ambiguous. For example, consider the question:

Non-specific question: "Do you like Mirasol?"

This question is too general. Is it asking if the respondent likes the hotel, the location, the bed linen, the pool, the restaurant and so on? If the hotel received a low score on this question, it would not know how to increase customer satisfaction. Better to know exactly what customers like and do not like to make the responses actionable.

Here is another example of a non-specific question:

Non-specific question: Do you stay at Mirasol regularly?"

Imagine that a respondent Sally says yes, since she stays every other year, without fail. Joe says no because he usually stays every month but missed last month. What does "regularly" mean?

Both examples can be rectified by asking:

Better: Rate the features of the following Mirasol recreational facilities:

[and then list features]

Better: How many times in past 3 years have you stayed at the Mirasol property?

(This avoids the "regularly" interpretation.)

There are several formats for the well-crafted questions that we can ask. For example, we can ask a simple yes/no question. These are called *dichotomous choices* that would be reported as percentages.

A multiple-choice question allows the respondent to choose the best answer from a list or to choose as many as apply. We can report a percentage response for each choice, or a Pareto chart by frequency of the choice categories.

Other questions might ask the customer to rank items. Respondents could be asked to choose the top three from a list or rank all the items in a list. Here we can report average ranks or create a Pareto chart.

For multiple choice questions that ask for a single response, a lot of surveys use what is called a Likert scale. The Likert responses are often treated as continuous data and are usually numbered from 1 - 5, 1 - 6, or 1 - 7. Many analysts are of the opinion that using more than seven points doesn't produce better information. Having too fine a scale, such as choices from 1 to 10, makes it difficult for respondents to consistently distinguish between scores of 4 and 5, or 7 and 8, for example. For these Likert scale questions, we can report an average score, or a percentage for each category.

Finally, there are open-ended questions. These might be best of all, since we allow the respondent to give reasons for their ratings, or to give feedback that touches on topics that haven't been covered in the survey questions. To analyze this type of data, we can use a tally sheet, a Pareto chart, or a verbal summary.

4.5 Analyzing Survey Data

An interesting analysis approach to survey data is to plot the responses of two variables on a perceptual map, as shown in Figure 4.7.

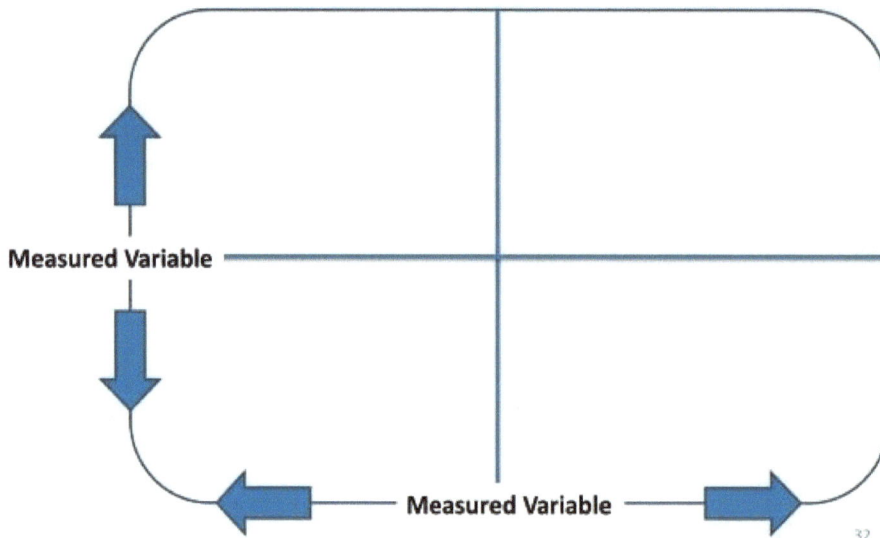

Figure 4.7 Perceptual map

Consider a survey that lists various activities, procedures, or amenities at the Mirasol Hotel, and for each, guests are asked how important that feature is (say, on a 1-5 Likert scale) and how satisfied they were with the feature (again, on a 1-5) scale. We can plot the paired responses for each item on a perceptual map as in Figure 4.8. This map shows that the ferry service scores high in satisfaction for the customers, but it is not that important to them. The restaurant is very

Customer Perceptions from SIR Survey

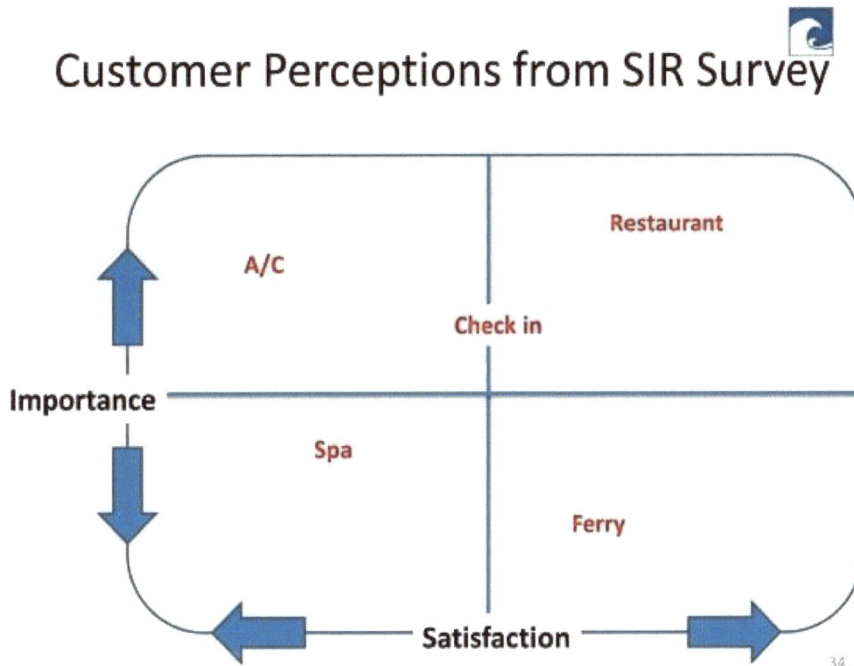

Figure 4.8 Customer perceptions from survey

important and has a high satisfaction rating. From this map, Mirasol can determine that it should concentrate on improving the air conditioning, since it is highly important, and the satisfaction rating is low. These types of graphical analyses go over well in meetings, because the items are placed in natural categories without a lot of numbers.

There is another type of survey that can be used to discover what is truly important to the customer in terms of Kano analysis: determining which features fall into the must-have category, which are delighters, and which are one-dimensional (also called satisfiers). Here, we ask questions about features of products or services and present them in pairs. The first time, we ask about the presence of the feature, and the second time, we ask about the absence of the feature. The presence question is also referred to as the functional question (the feature functions well) and the absence question is known as the dysfunctional question.

For example, Mirasol could develop a Kano survey on the amenities it offers at the hotel. An example *presence*, or *functional question* may be:

If the bedding in a hotel is high quality, how do you feel?

- o I like it that way
- o I expect it that way
- o I am neutral
- o I can live with it
- o I dislike it that way

Next, we have the absence or dysfunctional question to complete the pair:

If the bedding in a hotel is not high quality, how do you feel?

- o I like it that way
- o I expect it that way
- o I am neutral
- o I can live with it
- o I dislike it that way

After the results are in, each feature can be classified using the table in Figure 4.9 (Phillips, 2011) The M stands for must-have, O is one-dimensional (satisfier), and D is delighter. There are other categories on this table as well: the letter I is *indifferent*, meaning that the customer doesn't really have an opinion about the feature one way or another. The letter Q is *questionable*, indicating that the customer answered the pair of questions in an inconsistent manner. The letter R is a *reversal*, meaning that the presence of this feature would be a negative in the opinion of the customer.

If a customer answers that she likes the presence of a feature, but would expect not to have it, or is neutral or can live without it, then that feature is a *delighter*. If the customer likes the feature, but would dislike not having it, it is a *one-dimensional* or the more the better feature. If the guest expects, is neutral or can live with a feature, but would dislike it if it is not there, then that would be classified as a *must-have*. All other combinations fall into indifferent, reversal or questionable.

Classifying Kano Features

Customer Requirements		Dysfunctional				
		Like	Expect	Neutral	Live with	Dislike
F **u** **n** **c** **t** **i** **o** **n** **a** **l**	Like	Q	D	D	D	O
	Expect	R	I	I	I	M
	Neutral	R	I	I	I	M
	Live with	R	I	I	I	M
	Dislike	R	R	R	R	Q

M – Must Be
O - One Dimensional
D – Delighter

I – Indifferent, customer doesn't really care about the feature
Q – Questionable, customer answers are inconsistent
R – Reverse, features actually make customer feel worse

37

Figure 4.9 Kano classification table

How would these Kano survey results be used? The team would review the percentages of each Kano category for each feature. Features with the highest percentage of responses falling in the must-have and/or the one-dimensional category must be consistently delivered to customers. Those features that fall mostly in the indifferent category can be eliminated since those features do not affect customer satisfaction. Reversals should be eliminated, and delighters can be introduced at intervals to boost customer satisfaction.

Chapter 5 Choosing Metrics

Most companies have no shortage of data. The problem is shifting through it all and determining which variables can be used for decision making. Of the hundreds of metrics tracked on weekly reports, which are the ones that give actionable information?

As the saying goes, "What gets measured does get done," but sometimes, it's at the expense of other equally important tasks. Taking action based on a single variable has its drawbacks. Metrics must be chosen carefully so that a complete picture of the business is presented. Otherwise, we may take actions that sub-optimize a process.

For example, say a trucking company decides to measure the percentage of its on-time shipments. This metric becomes part of the weekly dashboard report, and managers are rated on how well they are meeting this metric. After a few weeks, the report shows that the percentage of on-time shipments is increasing, which is a positive thing, right?

An unintended consequence of stressing the on-time delivery measure, however, might be that to maximize on-time delivery, the dispatchers are now sending out trucks with half loads. While on-time shipments have increased, so did unused cargo capacity and fuel costs! The company needs to track both the on-time and capacity metrics to assure it is not sub-optimizing its process.

The trucking company should also look at the on-time delivery metric and ask some questions about what the customer really expects. What are the customers' true expectations for delivery? They are not all the same. Are we expending energy to deliver a product in a week when the customer would be just as happy if it got the shipment in two weeks? The trucking company needs to collect voice of the customer data for each of its accounts to define requirements more precisely. It needs to aim for fast and reliable service, but those definitions may differ for each customer.

5.1 Types of Metrics

There are three categories of key performance indicators (KPIs) (Eckerson, 2006). From the lowest to highest levels, they are operational, tactical, and strategic metrics.

The *operational* key performance indicators are measurements that are the closest to the actual work being performed. These metrics are out on the shop floor or the call center, for example, and are close to real time monitoring. Operational metrics can be used to trigger alerts and alarms.

The *tactical* KPIs provide a single version of the truth and are consolidated. They allow for proactive analysis and action. These metrics are what first-level managers rely on to track progress and make decisions.

Then, at the highest level, we have *strategic* metrics, such as financial or customer metrics that allow companies to make long range plans.

Within each category of KPI, there are two types of indicators: leading, which will predict future outcomes, and lagging, which are measurements of the past. We can influence a leading indicator, whereas a lagging indicator tells us how we did in the recent past.

Figure 5.1 shows some examples of leading versus lagging metrics (Eckerson, 2006). Leading metrics are predictive: for example, the number of clients the salespeople meet each week is predictive of future sales. Booked revenue is also a leading indicator. Monthly sales revenue itself is a lagging indicator because those sales transactions have already occurred.

The number of positive to negative customer comments can be a leading indicator of employee satisfaction or turnover.

Leading versus Lagging KPIs

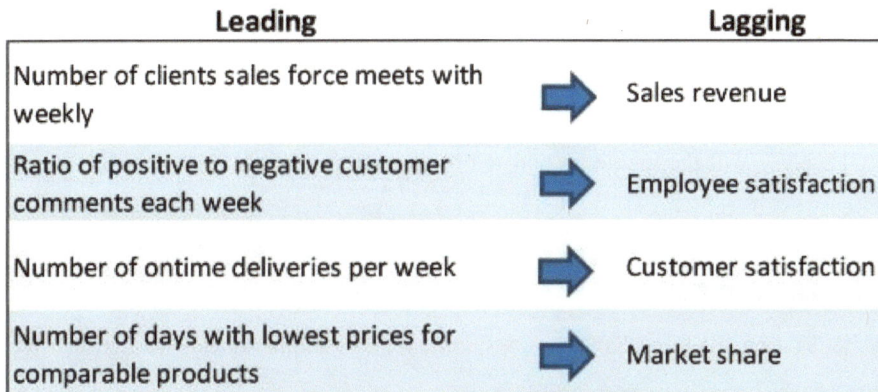

Leading		Lagging
Number of clients sales force meets with weekly	→	Sales revenue
Ratio of positive to negative customer comments each week	→	Employee satisfaction
Number of ontime deliveries per week	→	Customer satisfaction
Number of days with lowest prices for comparable products	→	Market share

8

Figure 5.1 Leading versus lagging KPIs

The number of on-time deliveries to a time sensitive client is a leading indicator, whereas customer satisfaction scores are lagging. The number of days in a month that a company has the comparable lowest price is a leading indicator of quarterly market share. Obviously, given the choice, we would like to have leading indicators so that we can adjust as we go.

5.2 Features of Good KPIs

How can we choose good KPIs for the dashboard report, or for our Six Sigma projects? The best KPIs are: (Eckerson, 2006)

- Aligned to the business's goals, or the project goals
- Owned in terms of data collection, analysis, and responsibility to act
- Predictive of future or of a broader truth
- Actionable – we must have the ability to improve the metric
- Few – we want to avoid dashboard overload

- Easy to understand – indices are sometimes too complex. When people can't understand what goes on under the hood in a formula, they won't be able to act to improve an index.
- Balanced and linked – this helps us avoid sub-optimizing a process
- Standardized – this will allow us to compare departments divisions, and cuts down on gaming the system by choosing different denominators in our rates, etc. We need one consistent version of the truth.
- Timely – will the metric be updated often enough for it to be actionable? Is the data easy to collect?

5.3 Choosing KPIs

Taking the features of a good KPI into consideration, how might we choose metrics? We must answer these questions (Eckerson, 2006):

- What is the business justification and strategic importance of this metric?
- Who is the metric's business owner?
- Who collects, reviews and reports data? Is this metric owned?
- What are the metric goals? (target and stretch)
- Who approves goals?
- How is the metric calculated? Is it clear? Standardized?
- What is the data source? How is it collected? When is data available? How easy is it to collect? How often will the metric be refreshed? How reliable is the data? Who must clean the data?
- Are there detailed reports to support the metric? Do we have a data trail?
- What are the related metrics (upstream and downstream)? We need to have our KPIs balanced and linked.

5.4 Quality Costs

Another type of process metric is the cost of quality, also referred to as the cost of poor quality. You may recall from Green Belt that there are four quality costs as identified by Joe Juran (Juran, 1980). The quality costs are internal and external failures costs, appraisal, and prevention.

Internal failure costs are the costs associated with mistakes the customer does not see. These are errors that are caught in the office, or on the factory floor before a product or service is delivered. Examples of internal failure costs include scrap, rework, disposition of

nonconforming material, overtime due to rework, and so on.

In contrast, *external failure costs* are the costs associated with mistakes the customer does see and experience. Examples of external failure costs include complaint adjustment, recalls, returns, warranty costs, and the loss of good will. This loss of good will is something that is hard to put a dollar figure on, but it is very real.

Appraisal costs are all the costs associated with inspection and testing. This includes overhead for a quality assurance department, calibration of test equipment, materials lost to destructive testing, and inspections. . Ideally, we would like to eliminate the need for inspection by making consistently making good quality. However, companies in regulated industries are often required to have quality assurance departments and are required to carry out specific inspection sampling plans. This is a cost that they can't avoid. For most companies, inspection is expensive.

Prevention costs are the costs associated with preventing defects. This includes training, visual management, mistake proofing devices, and process control. These are all proactive activities to prevent defects from occurring.

Of the four costs of quality, it is wise to spend the most on prevention than on the other three.

Let's examine a life cycle of quality costs. Refer to Figure 5.2. At first, a company would increase its inspection and try to improve its prevention activities. Appraisal and prevention costs increase. Since the company is inspecting more, initially, the internal failure costs increase because it is catching more mistakes. However, external failure costs decrease because the mistakes are caught before they get out the door and to the customer. This is desirable, but the company is not finished yet. Its overall costs are still too high.

At First, as Appraisal & Prevention Costs Increase...

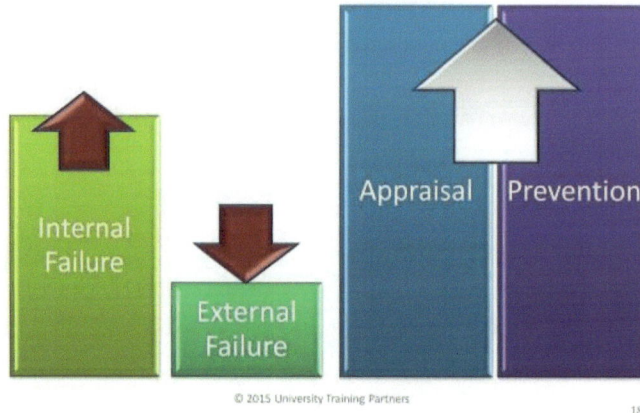

Internal Failure

External Failure

Appraisal | Prevention

© 2015 University Training Partners

18

Figure 5.2 Quality cost life cycle

Figure 5.3 illustrates that as prevention measures start to take hold, the internal failure costs also decrease, because mistakes are being prevented from occurring in the first place.

Then, Internal Failures Decrease...

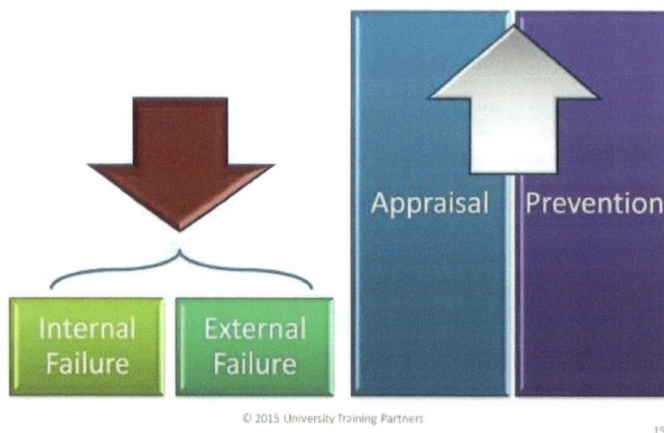

Appraisal | Prevention

Internal Failure | External Failure

© 2015 University Training Partners

19

Figure 5.3 Quality cost life cycle step 2

Then, eventually, the prevention activities such as mistake proofing, visual management, and training become the primary cost. (Figure 5.4) Because so many processes have been mistake-proofed, the company can rely less on inspection to assure quality, causing appraisal costs to

decrease as well. This is the desired end state.

Finally, Preventive Methods Reduce the Need for Appraisal

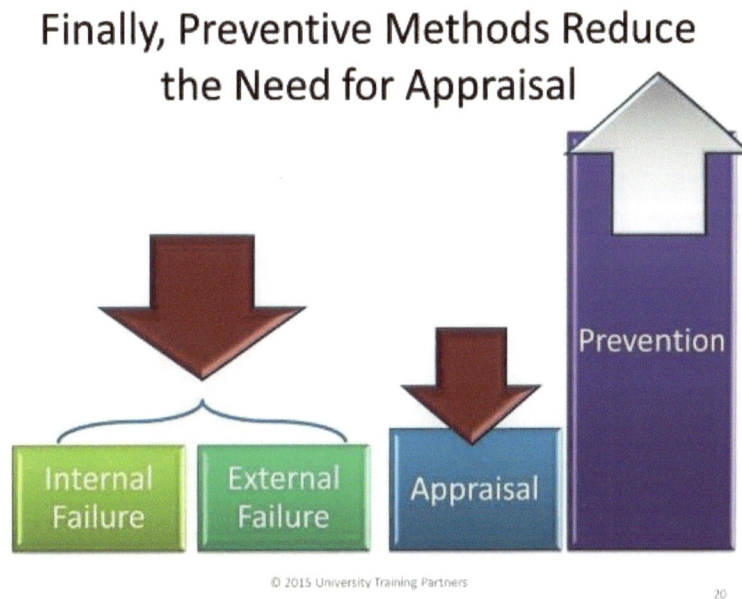

© 2015 University Training Partners

20

Figure 5.4 Quality cost life cycle end state

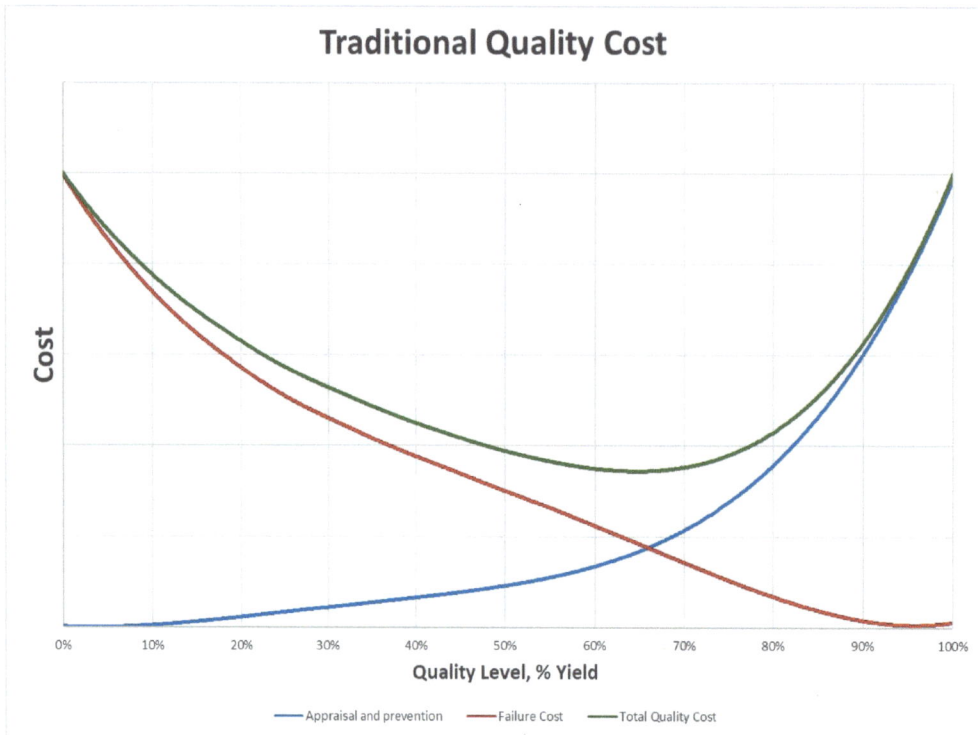

Traditional Quality Cost

Figure 5.5 Traditional quality cost

Traditionally, industrial engineering students were taught that a "sweet spot" existed the on total cost of quality curve. This minimum cost was achieved at the 65% quality level, as shown in green in Figure 5.5. The traditional model shows that internal and external failure costs (red line) decrease as the appraisal and prevention costs (blue line) increase. Because the mitigating effects of prevention are not considered in this model, the combined appraisal-and-prevention line was mostly based on costly inspections. Accordingly, as the quality level increases past the 65% mark, the appraisal cost becomes prohibitive with the curve rapidly increasing as the quality level approaches 100%. This type of quality cost graph is typical of the thinking of the US in the 1950s.

In the modern quality cost graph shown in Figure 5.6, the lowest total quality cost is achieved at 100% quality. At the 100% point, not only have we eliminated the internal and external failure costs, but the positive effects of prevention overcome the need for appraisal. This mirrors the quality mindset of Japanese manufacturers in the 1950s.

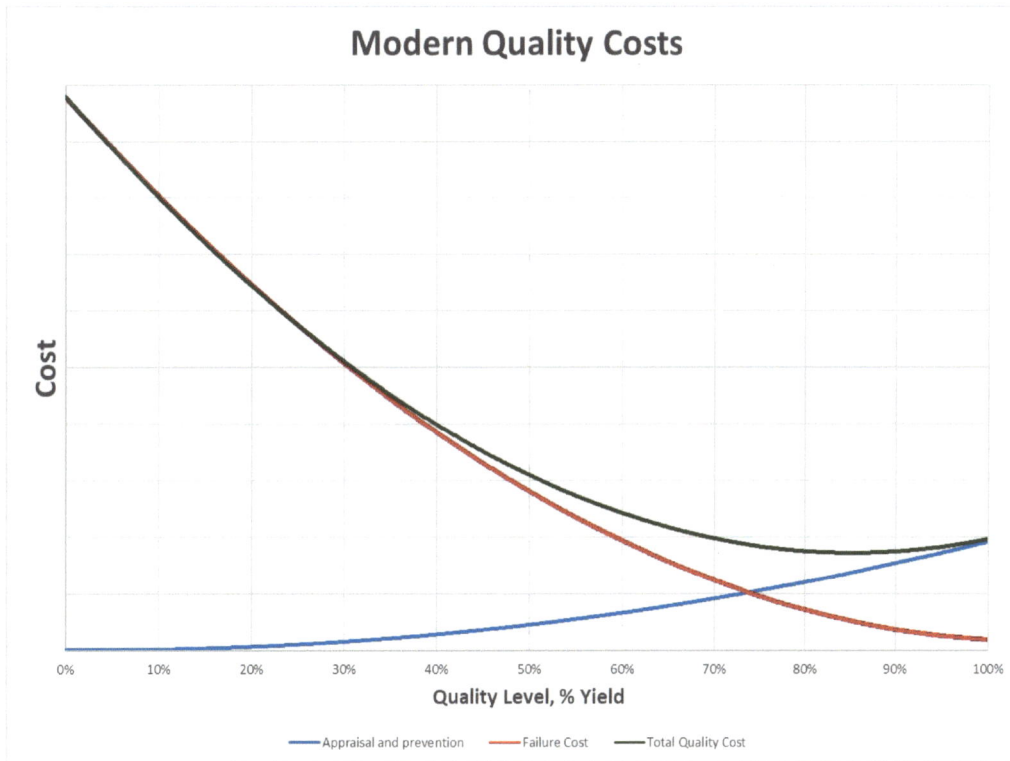

Figure 5.6 Modern quality cost

5.5 Financial Metrics

Because Six Sigma projects stress bottom-line results, it is important for Black Belts to understand basic financial metrics. The first term we will consider is *margin*. Margin is equal to revenue minus cost, or profit. We can also calculate a percent margin to compare the margins across projects.

$$\text{Margin} = \text{Revenue} - \text{Cost}$$

$$\text{Percent Margin} = [(\text{Revenue} - \text{Cost}) / \text{Revenue}] \times 100\%$$

If we want to know what the return on investment is for a short-term project, we can use the *Simple ROI* formula. This formula ignores the time value of money and can be used for projects in which the effect of compound interest is negligible, such as projects which are less than a year in duration.

To calculate Simple ROI, we take the financial benefits of the project, which could be dollar savings, and/or an increase in revenue or profit, and subtract off the cost of the project to yield net benefit. Then we divide the net benefit by the project cost. The resulting Simple ROI gives us an idea of the benefit of the project's monetary outlay.

$$\text{Simple ROI} = (\text{Benefit} - \text{Cost}) / \text{Cost}$$

For example, if a Six Sigma improvement project costs $55,000 to implement, but it produces a one-time increase in sales or an improvement in delivery time equal to $100,000, then the Simple ROI is calculated as:

$$\text{Simple ROI} = (100{,}000 - 55{,}000) / 55{,}000 = 0.81.$$

This is an 81% return on investment, meaning that we recouped not only the entire cost of $55,000, but we also brought in 81% of the cost in addition.

A positive ROI indicates that we have made money on the project. An ROI of zero means that we have broken even. A negative ROI indicates that the benefits of the project were not enough to recoup the cost of the project.

More sophisticated financial analyses consider the time value of money. If a benefit will be received in the future, it must be brought back to today's dollars using a *net present value* calculation so we can properly evaluate its merits. For example, imagine you were told that you would receive $20,000, but the payment would be made 20 years from now. You can calculate the *net present value* to find what that future $20,000 benefit would be in today's dollars based on the inflation rate. Will it be enough to go on a European vacation?

The *net present value* (NPV) is calculated:

$$NPV = \frac{A}{(1 + r)^n}$$

The variable A is the payment you will receive, $20,000. The variable r is the rate. Here, let's assume the rate of inflation is 2%. We turn 2% into a decimal and write it as 0.02 in the formula. The variable n is the number of years you will have to wait for the payment. As each year goes on, the buying power of that $20,000 is slowly eroding. The net present value is equal to A divided by 1 plus r to the nth power. We can put this into our calculators to get $13,459.43, which is much less than the payment's buying power today.

$$NPV = \frac{\$20,000}{(1 + 0.02)^{20}} = \$13,459.43$$

Alternatively, we can use the *=npv* formula in Excel, shown in Figure 5.7. The arguments for this formula include the rate, and Value1, which is an array of the value of payments received. In this case, the payments for the first 19 years are equal to zero, and the 20th payment is equal to $20,000. Create a column in Excel with 19 zeroes followed by a cell with 20,000. Use this cell range for Value 1. The formula will return the net present value.

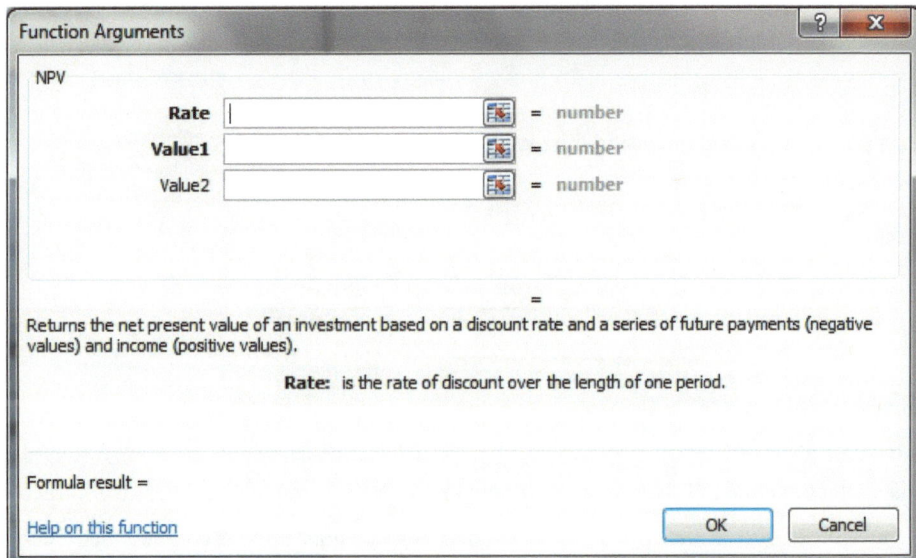

Figure 5.7 Excel NPV formula

Let's work through another example. In ten years, a company will receive a payment of $350,000. Assuming an inflation rate of 3%, what is the NPV? The calculation follows.

NPV = ?, A = $350,000,   r = 0.03, n = 10

$$NPV = \frac{A}{(1 + r)^{10}}$$

$$= \frac{\$350{,}000}{(1 + 0.03)^{10}} = \$260{,}432.90$$

Note the formula we used for NPV assumes that money is compounded once a year. We can use a more general formula to consider that there may be an annual rate of return r, but the interest might be compounded monthly, quarterly, etc.

We now introduce a new variable, t, which is the number of times that the interest is compounded annually. If we let t equal 1, we get the first NPV formula that we used. This new, more general formula can be used in all situations.

$$NPV = \frac{A}{(1 + r/t)^{10t}}$$

Let's do the previous example using the more general formula:

NPV = ? A = $350,000, r = 0.03, t = 1, n = 10

$$NPV = \frac{A}{(1 + r/t)^{10t}} = \frac{\$350{,}000}{(1 + 0.03)^{10}}$$

$$= \$260{,}432.90$$

Here is another example of the use of the more general NPV formula. A company will receive a payment of $125,000 in three years. The interest rate is equal to 5%, compounded every six months (twice a year). Here, A = $125000, r = 0.05, t = 2 and n = 3. What is the net present value of the payment?

NPV = ? A = $125,000, r = 0.05, t = 2, n = 3

$$NPV = \frac{A}{(1 + r/t)^{nt}}$$

$$= \frac{\$125{,}000}{\left(1 + 0.05/2\right)^{3 \times 2}} = \frac{\$125{,}000}{(1 + 0.025)^{6}} = \$107{,}787.10$$

If on the other hand, the interest is *continuously* compounded, the NPV formula converges to:

$$NPV = \frac{A}{e^{rn}}$$

where e is the constant 2.718...

For example, if a payment of \$100 will be paid in 5 years, and the interest rate is 1%, continuously compounded, the net present value is:

NPV = ? A = \$100, r = 0.01, n = 5

$$NPV = \frac{A}{e^{rn}} = \frac{\$100}{e^{0.01 \times 5}} = \$95.12$$

We can also find the NPV for a series of payments over several years as well using the formula:

$$NPV = \sum_{n=0}^{n} \frac{A_t}{(1+r)^n}$$

For example, let's calculate the NPV of a Six Sigma project that requires an initial outlay of \$10,500 and delivers savings of \$4500 in the first year, and \$8900 in the second year. Required rate of return is 8.0%, compounded annually.

To answer this question, we first need to write out the cash flow by year, and then apply the NPV formula.

NPV = ?, r = 8%, n = 2

Yr Cash Flow
0   -\$10,500
1   \$4,500
2   \$8,900

$$NPV = \sum_{n=0}^{n} \frac{A_t}{(1+r)^n}$$

$$= \frac{-\$10,500}{(1+0.08)^0} + \frac{\$4,500}{(1+0.08)^1} + \frac{\$8,900}{(1+0.08)^2}$$

$$= -\$10,500 + \$4166.667 + \$7630.316 = \$1296.98$$

Remember to round the answer at the end, not in the intermediate calculations.

Using the NPV, we can calculate a *discounted ROI* that takes the time value of money into account.

Discounted ROI = (Total PV Benefits – Total PV Cost) / (Total PV Cost)

Here, benefits in future years are brought back to the present as are all the future costs. The formula for discounted ROI is:

$$Disc\ ROI = \frac{\sum_{i=0}^{n} \frac{Benefit_i - Cost_i}{(1+r)^i}}{\sum_{i=0}^{n} \frac{Cost_i}{(1+r)^i}}$$

The use of the formula is illustrated in the following example.

A company is considering leasing a new piece of equipment with a two-year contract. The initial cost is $35,000 with maintenance fees of $4,000 for the next two years. The immediate benefit is equal to $50,000 and $5,000 for each of the next two years. What is the overall discounted ROI? Assume the discount rate is 10.0%.

We first draw the cash flow table:

Yr	Cost	Benefit
0	$35,000	$50,000
1	$4,000	$5,000
2	$4,000	$5,000

And then write out the values for r and n:

$r = 0.10, n = 2$

Next, we plug the values into the discounted ROI formula:

$$Disc\ ROI = \frac{\sum_{i=0}^{n} \dfrac{Benefit_i - Cost_i}{(1+r)^i}}{\sum_{i=0}^{n} \dfrac{Cost_i}{(1+r)^i}}$$

$$Disc\ ROI = \frac{\dfrac{50{,}000 - 35{,}000}{(1+0.10)^0} + \dfrac{5{,}000 - 4{,}000}{(1+0.10)^1} + \dfrac{5{,}000 - 4{,}000}{(1+0.10)^2}}{\dfrac{35{,}000}{(1+0.10)^0} + \dfrac{4{,}000}{(1+0.10)^1} + \dfrac{4{,}000}{(1+0.10)^2}}$$

$$Disc\ ROI = \frac{\$16{,}735.54}{\$41{,}942.15} = 0.40$$

We will realize a return on investment equal to 40%.

Chapter 6 Leading Teams

When we first begin a project, before the team is chosen, we perform a vital exercise called a stakeholder analysis. From this analysis, we will identify stakeholders and develop a plan to bring them onboard.

6.1 Stakeholder Analysis

A stakeholder is anyone who has a personal stake in an issue or who can create or be affected by an action or change. We can identify stakeholders by asking these questions. The answers will be stakeholders. For our project, we ask:

- Who might receive benefits or experience negative effects?
- Who might be forced to make changes or change behavior?
- Who has goals that align/conflict with this project?
- Who has responsibility for action or decisions?
- Who has resources or knowledge that is important to the project?
- Who has expectations for this project?

All the departments or people identified by the answers to these questions will be stakeholders.

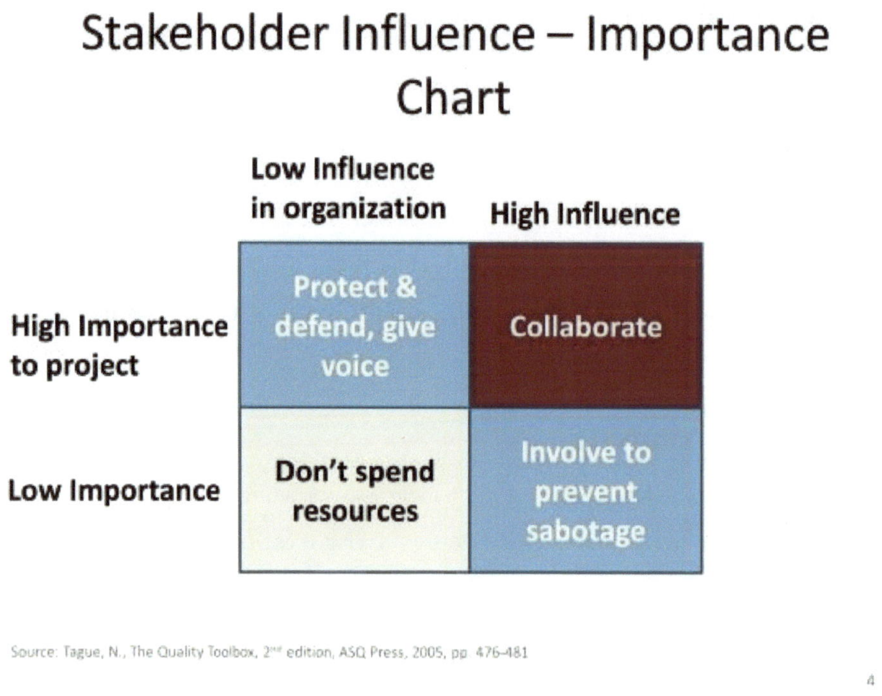

Stakeholder Influence – Importance Chart

	Low Influence in organization	**High Influence**
High Importance to project	Protect & defend, give voice	Collaborate
Low Importance	Don't spend resources	Involve to prevent sabotage

Source: Tague, N., The Quality Toolbox, 2ⁿᵈ edition, ASQ Press, 2005, pg. 476-481

4

Figure 6.1 Stakeholder influence-importance chart

After specific stakeholders are identified, each is placed into one of four quadrants on an Influence & Importance chart, as shown in Figure 6.1 (Tague, 2005). Here, *influence* refers to influence in the company, and *importance* means importance to the project. Each quadrant of the chart gives a recommended behavior approach based on a stakeholder's influence and importance. The team leader can then develop a targeted action plan to help each stakeholder accept the project and the changes it will bring.

If a stakeholder is very important to the project's success, but he or she has a low influence in the organization, then the behavior approach is to protect and defend the stakeholder and give them voice. For example, if a Six Sigma project involves redesigning a manufacturing process, the line workers would be highly important to the project success, but they would not have a lot of influence in the organization. We need to make sure these people will not lose their jobs because of the project, and we need to make sure they know this.

We also will need to give voice to the stakeholders in this quadrant by inviting them to team meetings or having them on the core team. Note that workers who have not previously been given a voice may have years of pent-up frustration aimed at management or the organization. It may be wise to have a pre-meeting with these team members so that they can express their concerns before the larger team convenes. This approach will increase the productivity of the first few team meetings.

The next behavior approach category is for stakeholders who are highly important to the project and have a high influence in the organization. This would include process owners, for example. We want to collaborate with these folks.

The stakeholders falling into the low importance and low influence category might be stakeholders, but their buy-in is not critical to the project's success.

Last, we have folks who are not important to the project per se but are highly influential in the organization. These stakeholders might control financial resources, or they may be upper-level administrators. The suggested behavior approach is to involve them to prevent later sabotage of the project.

6.2 Participation Plan

After identifying stakeholders and placing them on the Importance & Influence chart, we can next develop a participation plan, shown in Figure 6.2. (Tague, 2005) Depending on the level of change or commitment required of a stakeholder, he or she would be placed on the participation continuum.

Operating Definitions for Participation Levels

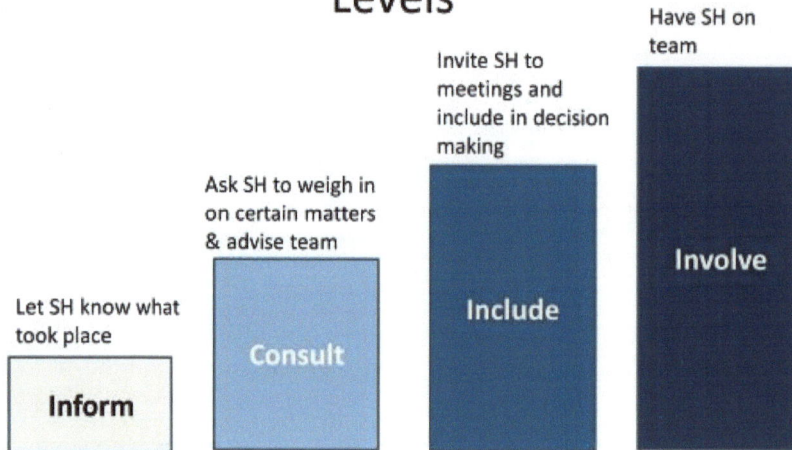

Have SH on team

Invite SH to meetings and include in decision making

Ask SH to weigh in on certain matters & advise team

Let SH know what took place

Involve

Include

Consult

Inform

6

Figure 6.2 Participation plan

Stakeholders who will not need to change very much are placed on the lowest level. We can decide to inform these stakeholders, letting them know what took place. We will send them meeting minutes or speak to them informally. For stakeholders who will experience a higher level of change or required commitment, we may choose to assign them to the Consult participation level. Stakeholders at this level will be asked to weigh in on certain matters. We will invite them to a meeting for which their expertise will be needed. As the change required increases even more, stakeholders are placed in the Include level, which may mean being regularly invited to meetings and being included in decision making. For those stakeholders required to undergo the most change, we will put them in the Involve level, which means putting them on the core team.

The stakeholder template in the course spreadsheet shown in Figure 6.3 lets the project leader name stakeholders and select one or more reasons why they are stakeholders. Then each stakeholder is placed then into Importance and Influence quadrants, and the leader chooses a participation level. Finally, the team leader selects a communication plan: will this stakeholder be sent a copy of minutes, will the stakeholder be spoken with regularly, and so on. As a last step, the leader assigns a person — by name — who will be responsible for delivering that communication.

The team lead can perform the stakeholder analysis with the project champion before the project officially begins. The analysis will help identify who should be on the team and who might need to be pulled in for some expert help. In addition, the analysis will provide direction on how to handle the stakeholders to assure the project goes as smoothly as possible. Dealing with people, personalities and office politics is often the most difficult part of leading a project.

6.3 Choosing the Team

There should only be a total of five to seven core members on a Six Sigma team. Having a larger team will impede progress and introduce unnecessary complexity. Of course, there may be other people the team brings in along the way – ad hoc team members – who share their expertise, but they will be temporary members who are invited to a few meetings to give their advice and opinions.

There are specific roles for each team member. For example, a champion, or sponsor, is a high-level manager that is not involved in the day-to-day team activities, but is available to acquire or free up resources, or remove roadblocks for the team.

Team leaders, as stated earlier, will be either a Green Belt or a Black Belt. It would make sense if the leader were also the process owner, but if not, the process owner would be another team member.

A facilitator is not a process expert but is there to make sure the team functions well and stays on target.

The scribe takes meeting minutes, keeps track of deliverables and due dates, and transcribes the team's flipchart and white board work. Try to make the scribe's job easy: don't make this thankless task too onerous.

Team members must attend meetings to lend their expertise. It is vital that members meet their deadlines for deliverables - the Six Sigma project will only be two to nine months in duration.

Stakeholder	Why a Stakeholder?						
	Will receive benefits/ negative effects	Will be forced to make changes	Has goals that align/ conflict with project	Has responsibility for action/ decisions	Controls important resources	Has useful expertise	Has expectations for project

Influence/Importance		Participation Level			
Influence in organization Lo/Hi	Importance to project success Lo/Hi	Inform	Consult	Include	Involve

Communication Plan					
Speak with informally as needed	Send copy of meeting minutes	Invite to team meetings	Meet with regularly	Other (describe)	Responsible for communication

Figure 6.3 Stakeholder template

6.4 Launching the Team

As project leader, it will be your responsibility to launch the Six Sigma project team. This team will most likely be cross-functional and have members from various levels of the organization. As such, the team members might not know each other very well, and they might not sit in meetings together very often, if ever.

In the first few meetings, it is the leader's responsibility to take this diverse group of people and turn them into a team working toward a single goal. Make sure to incorporate a team building or ice-breaking exercise in the kick-off meeting agenda to allow everyone to know each other. Ask the executive sponsor to attend the kickoff as well to emphasize the importance of the project and the value the organization places on Six Sigma. The project and goal statement and any relevant data from the project charter can be presented. In addition, lay out the project schedule and make sure everyone understands the time commitment associated with being a team member.

A useful exercise for the first meeting is for the group to create a team charter. This team charter exercise will help the team agree on the ground rules of the team and will make sure everyone has the same expectations. This will help break the ice and give the team something concrete to work on in the first meeting.

There is team charter template in the companion spreadsheet (Figure 6.4). The template is a jumping off point for the team. It lists expectations in the first column, and a "straw man" example of what the team's rule could be in the second column. It is easier to edit than to create, so the team can start with the suggested example and then tweak it to suit its needs. At the end, everyone on the team will sign the team charter. If there are some lower levels employees on the team, you might also want their supervisors to read and sign the document. This allows the supervisors to acknowledge that their employee has other time and meeting commitments in addition to his or her regular job.

Some examples of ground rules that come out of the team charter might include (Eckes, 2003):

1. Start on time and finish on time
2. One person speaks at a time
3. Limit sidebars
4. No stripes in the room

5. Everyone participates
6. Cell phones on vibrate
7. No acronyms without explaining first

Team Charter

Project Title	

Expectation	Example	Team Rule
Attendance	Attendance is required at all team meetings. Changes in meeting times must be made at least 24 hours ahead of time.	
Participation	Team members may not be substituted unless approved by team leader.	
Focus	We will stay on task and on topic, using the Project Charter as our guide. A meeting agenda will be publishedat least one day in advance.	
Interruptions	Interruptions for emergencies only. Phones turned to silent.	
Preparation	All deliverables are expected to be completed in a timely manner. Each meeting will have a published agenda.	
Timeliness	Meetings will begin promptly as scheduled.	
Decisions	We will choose the best decision-making method for each situation. We will support decisions made by the team.	
Data	We will rely on data to make decisions.	
Conflict	We welcome honest disagreements, as long as everyone is treated with respect. A facilitator will be used if conflict cannot be resolved.	
Other		

Team Member	Role	Signature

Figure 6.4 Team charter template

Recognize that in the first team meeting, some members will want to skip some of the DMAIC steps. If you bring seven people into a room to discuss a problem, what would they instinctively want to do? Solve the problem. The natural inclination is to skip Define, Measure, and Analyze and jump right into Improve. This is a positive sign, because it tells you that the team is enthusiastic about the project. But it is your job as team leader to pull the reins back on the team and tell them about the DMAIC cycle. If the problem were that easy to solve, it would have been fixed already. Everyone has an opinion on what needs to be done, but with no data, it is just gut feel at this point. Capture all the premature improvement ideas on a flip chart and save them. As team lead, you do not want to shut people down early and have them lose enthusiasm. You will need to gently lead them through the rationale behind DMAIC and discuss the reasoning behind each step.

6.5 Team Growth Stages

Any cross-functional Six Sigma project team will inevitably go through the standard team growth stages of forming, storing, norming, performing and finally adjourning with recognition. As a leader, you need to recognize each stage, and try to bring the team along the curve as quickly as possible (Figure 6.5).

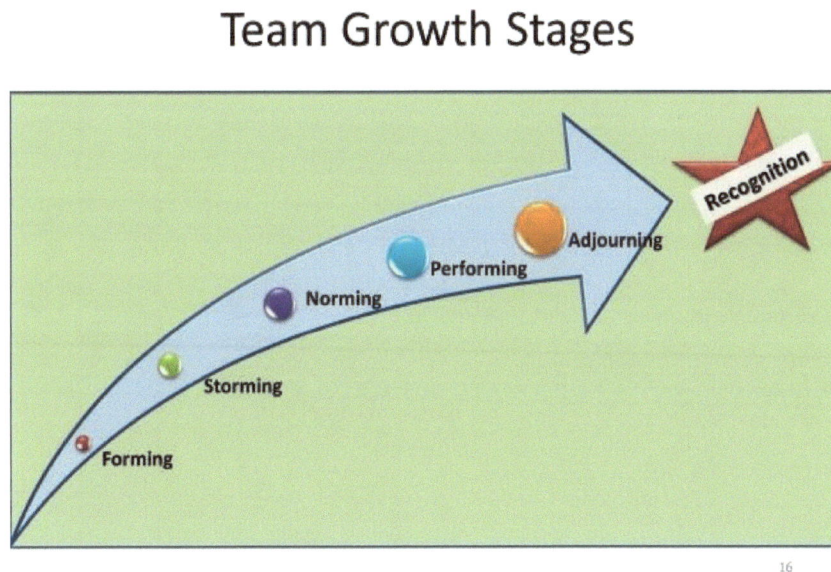

Figure 6.5 Team growth stages

At the *forming stage*, team members feel unsure, and are trying to figure out where they fit in, how they will contribute to the team. There may be some intimidation if there are line workers and engineers in the room, or higher-level managers.

In the *storming* stage, members are starting to be comfortable enough to disagree. At this stage, deep-seated resentment against management might surface, tension between departments might display itself, or a long-standing feud between production engineers and design engineers might bubble up to the top, and so on. In the storming stage, members are thinking of themselves as individuals and do not understand that they are all part of a unit. Recognize that this stage is a normal growth stage, but make sure that everyone is respectful when they disagree. The team will make it through to the other side.

After the storming phase comes the *norming* stage, during which the team begins to coalesce. Next comes the most productive and enjoyable stage which is *performing*. This is when the team strongly identifies as a single unit, and members work together to achieve the Six Sigma project goals.

There is a last stage, *adjourning*, which we add to remind us that the Six Sigma team is informal and will only last from two to nine months. To adjourn well, recognize everyone's hard work and contribution, and make sure upper management also acknowledges the team. Make sure all loose ends are wrapped up and the project is handed over to the process owners cleanly.

6.6 Team Behavioral Pitfalls

To perform effectively, all team members need roles and responsibilities. The team needs common goals, well-run meetings with agendas, ways to reach agreement, and ground rules. Even with these tools in hand, a team leader may still encounter problems with member behavior. Team members may not be prepared for meetings, they may miss deliverable deadlines, be late to meetings, or miss meetings completely. There must be consequences for poor team behaviors. Team peer pressure will not work especially well until the team has reached the norming development stage. Before then, the team charter can serve to achieve consensus on the expectations for members while the team is in the forming and storming stages.

Unfortunately, there are many ways that a team can fail in its goals. Knowing the potential pitfalls ahead of time can help the team leader nurture a positive team environment that produces results. Lencioni (2012) has identified the Five Dysfunctions of a team. To achieve results, a team must overcome these five dysfunctions: absence of trust, fear of conflict, lack of commitment, avoidance of accountability and inattention to results. These problems form a hierarchy as shown in Figure 6.6.

Figure 6.6 Five dysfunctions of a team

The five dysfunctions build upon each other. For example, trust is the foundation for the team. Without it, none of the other dysfunctions can be overcome. If members do not feel comfortable voicing their opinions, or making mistakes, then the team will not be successful.

Only after trust is established can members feel secure enough to disagree with each other. Trusted team members can challenge each other's views without creating resentment. If there is a fear of conflict, then the team will lapse into groupthink, which rarely results in the best solutions.

Teams that engage in healthy discussions foster commitment from members. The decisions made because of these discussions, in which everyone's opinion is heard, are more likely to earn buy-in from members.

Members who are committed to the team's decisions will not avoid being held accountable for their actions. Members' goals will align with the team's decisions, and they will work to achieve the desired result.

Finally, team members that trust each other, are not afraid of conflict, are committed to team decisions and are accountable to the group can pay attention to results, or the team's success.

There may be individual team members who exhibit behaviors that are not productive (Eckes, 2003). As a team leader, you will need a strategy to address each one of these behaviors, presented by Eckes as separate characters: the whisperer, the storyteller, the dropout, the naysayer, the verbal attacker, the politician, and the clown.

The *whisperer* who turns and has private conversations with the person next to them. Whatever he or she is whispering is automatically thought of as negative by the other members of the group.

The *storyteller* is well liked, but gets off on tangents, or must recount the entire history of a problem, often with themselves as the main character or hero. How do you rein someone like that in? There are always dominant personalities in every room: how do you elicit input from the more introverted members of the group? Is the dominant personality intimidating?

On the other end of the spectrum, we have the *dropout* who is not the least bit engaged. This team member may have been on the losing side of a previous decision made by the team and has since decided that he will not cooperate. When there are only five to seven team members and a nine-month timeframe, every member needs to be engaged.

The *naysayer* is a very common personality. A naysayer be the team member who has been at the company for many years, and who always chimes in that a solution was tried 20 years ago and it didn't work. Or it may be the team member who says no to every idea because it brings them a sense of being more knowledgeable than the other members of the group. How do you prevent the naysayer from shutting down the conversation?

The *verbal attacker* type of personality should not be tolerated. How have you dealt with this type of person before?

There is the *politician*, who sees every decision to get ahead, or the type who agrees with whoever they are currently talking to.

Finally, we have the team *clown*, who wastes a lot of time in meetings with asides and jokes. Comic relief is often welcome, but it has its limits. Is the clown productive? Is he meeting deadlines, or is he using humor to deflect attention away from his lack of accomplishments?

When there is a disruptive personality in a group, or in a presentation audience, a leader can use various actions following an escalation scale. (Eckes, 2003) At the lowest level, a leader can make eye contact with the offender. That often is all the leader needs to do to make someone aware of their behavior. If that approach fails, the leader can use other means such as standing up and walking halfway to the offender or walking by the person and making eye contact. A leader may choose to bring attention to the disruptor by asking what he or she thinks about what is being said. This is an effective technique if the team member continues to whisper or sidebar, or if they have dropped out. If these approaches do not bring the desired result, the leader can talk to the offender directly, or confront them on a break. At the most extreme level, a leader could choose to confront the person directly in front of the whole group. Recognize of course that there are a lot of actions that can be taken before resorting to public confrontation.

6.7 Team Decision Making

How can teams make decisions? As shown in Figure 6.7, there is a continuum of methods starting with authoritarian, consultative, consensus, 2/3 majority and 100% agreement. (Eckes, 2003) Each has its place, and each has its advantages and drawbacks. The best way forward depends on the team and the decision at hand.

Team Decision Making Approaches

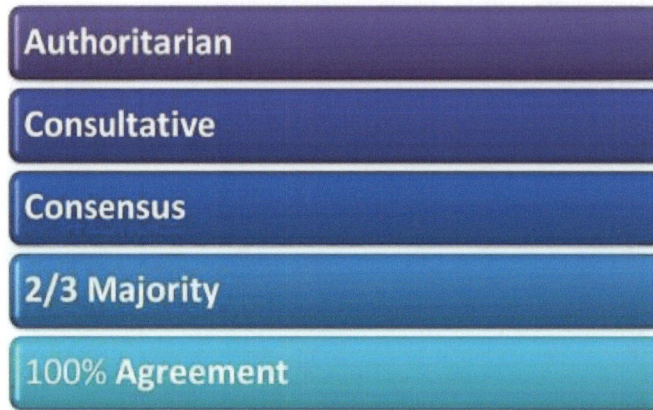

Authoritarian

Consultative

Consensus

2/3 Majority

100% Agreement

Source: Eckes, Six Sigma Team Dynamics, (2003) p. 58

21

Figure 6.7 Team decision making

Authoritarian decisions are quick since the team leader just tells everyone what to do. Of course, this is not the best way to build teamwork. Some decisions might need to be made this way, but overall, it is not in keeping with the idea of bringing a cross-functional group together to improve a process.

In consultative decision making, the leader asks each member his or her opinion before deciding. Here, the team members are asked for input, but in the end, the decision is made by the leader.

In a consensus decision approach, everyone's opinion is sought, and the team decides together what action to take. Sometimes, though, in its effort to appease everyone, the team's final decision is watered down and not as effective as it could be.

Using majority vote is the democratic way, I suppose, and the decisions do not have to be watered down. However, within the confines of the small team, this method creates winners and

Chapter 6 Leading Teams © 2021 University Training Partners

losers which may cause bad feelings among team members. Those on the losing side may become less engaged with the team going forward,

Using the 100% Agreement, or unanimous decision, is slow and difficult to achieve. As such, it is not recommended.

6.8 Facilitation Techniques

A team leader must get the most out his or her team members. The core team members were carefully chosen through the stakeholder analysis, and each has an important perspective and skill set to add. Knowing how to facilitate discussions and draw even the introverted team members out is an important skill. There are several facilitation techniques that can be used to lead group discussions (Eckes, 2003). Among these are the direct probe, the re-direct, the confirmatory statement, the leading question, the behavioral observation, idea floating, the boomerang and the lasso.

A leader can ask a direct question, or use a *direct probe*, such as, "What behaviors are unacceptable for our team?" Or "How might we work best as a team?"

The *re-direct* method can be used to bring the conversation back to the topic at hand in case it gets off track. For example, "Let's talk about the team behaviors we have listed on the board."

An example of a *confirmatory statement* is, "I'm hearing that you want to meet only in the afternoons, and others have said that mornings are too busy, is this correct?"

A facilitator can ask *leading questions* that will force a response or a change. For example, the leader can challenge a sluggish group and ask, "I thought this group would come up with at least 20 improvement ideas. What else can we do to improve this process?"

A *behavioral observation* might take the form, "I see that many of you have sunk down in your seats once we brought up this topic. This must be an uncomfortable issue."

Idea floating is throwing out a hypothetical solution just to get the conversation going, such as "What if we took all the machines and rearranged them in a circle? Would that work?"

The *boomerang* is to take a person's question or comment and hand it back to them. For example, if a team member, Joe, comments, "That idea won't work," turn it around and ask,

"Joe, what approach do you recommend?"

Using the *lasso* technique, the facilitator gathers all the group's thoughts and makes an overall, summary statement.

Being a facilitator is not an easy task. Most of us are thrust in team leader positions without adequate training on running meeting and guiding teams. Knowing what facilitation behaviors to avoid can help a team leader evaluate his or her performance and make corrections going forward. Eckes (2003) lists ten sins of facilitation. These are committed when the facilitator:

1. Chooses which comments are worthy of being documented. This is the best way to shut down a team and have people disengage!
2. Shows bias toward one tool or technique. There are many tools in a Black Belt's toolbox. Don't fall in love with two or three.
3. Permits digressions, which means nothing gets done.
4. Permits ground rules to be broken. Once the rules are broken, no one respects the rules going forward.
5. Creates the impression that he or she has a bias toward one person or idea. A facilitator should be dispassionate about the ideas presented.
6. Speaks in emotionally charged language.
7. Allows distrust or disrespect to occur.
8. Doesn't create a sense of purpose.
9. Ignores time keeping.
10. Does the work for the team.

Chapter 7 Managing Projects

7.1 Tollgate Checklists

In the companion spreadsheet, there are a series of tollgate checklists that lists the most probable activities performed in each of the five DMAIC phases. (Figures 7.1 – 7.5). The checklists will be useful when preparing documentation for the tollgate meetings with the project champion. Meetings will be held after each project phase. The team will receive clearance to move to the next stage after the champion is signs off on the progress up to that point (Pande, 2002).

Define Phase Checklist

Check	Activities	Comments
	Sections 1-9 of the project charter have been reviewed by team and changes made as necessary.	
	SIPOC completed.	
	Team charter has been drafted and signed by team members.	
	VOC data gathered (Kano analysis)	
	Preliminary CTQs defined and linked to KPOVs.	
	Stakeholder Analysis completed.	
	Communication and Participation Plan in place.	
	Gantt chart created.	
	Tollgate review prep	
	Tollgate review conducted & approved.	Date:

Figure 7.1 Define phase checklist.

Measure Phase Checklist		
Check	**Activities**	**Comments**
	Detailed process map created.	
	KPOVs and KPIVs identified.	
	Measurement system analyzed.	
	Data collection plans created that include sample size, frequency, responsible party and duration of data collection.	
	Baseline analysis performed.	
	Updated project charter and plans.	
	List of quick improvements created and parking lot items preserved.	
	Tollgate review prep	
	Tollgate review conducted & approved.	Date:

Figure 7.2 Measure phase checklist

Analyze Phase Checklist		
Check	**Activities**	**Comments**
	Process analysis performed.	
	Root cause analysis performed.	
	Critical KPIVs identiied.	
	Updated project charter and plans.	
	Tollgate review prep	
	Tollgate review conducted & approved.	Date:

Figure 7.3 Analyze phase checklist

Pre-Pilot Improve Phase Checklist

Check	Activities	Comments
	List of innovative potential solutions generated.	
	Candidate improvement selected for pilot implementation.	
	Pilot and pilot control plan developed.	
	Pilot presentation prep	
	Pilot review conducted & approved.	Date:

Post-Pilot Improve Phase Checklist

Check	Activities	Comments
	Pilot is conducted.	
	Pilot results evaluated.	
	Solutions refined based on lessons learned from pilot.	
	Plan developed to expanded plot to full-scale implementation.	
	Updated project charter and plans.	
	Tollgate review prep	
	Tollgate review conducted & approved.	Date:

Figure 7.4 Improve phase checklists

Control Phase Checklist		
Check	**Activities**	**Comments**
	Full-scale implementation results compiled.	
	Confirmation that project goals achieved.	
	Documentation and measures for control plan developed and implemented.	
	"Lessons Learned" developed and communicated	
	Reward and Recognition scheduled.	
	Process and control plan turned over to process owners.	
	Tollgate review prep	
	Tollgate review conducted & approved, project closed.	Date:

Figure 7.5 Control phase checklist

7.2 The Gantt Chart

The activities of the project itself or, for each phase of the project, can be tracked using a Gantt chart (Figure 7.6). Gantt charts list specific tasks, the person or department responsible for each task, the start and end dates, and the duration of each task. Percent completion is tracked daily. Software such as Microsoft Project can be used to track project progress, but for smaller projects, a spreadsheet might be all that is needed.

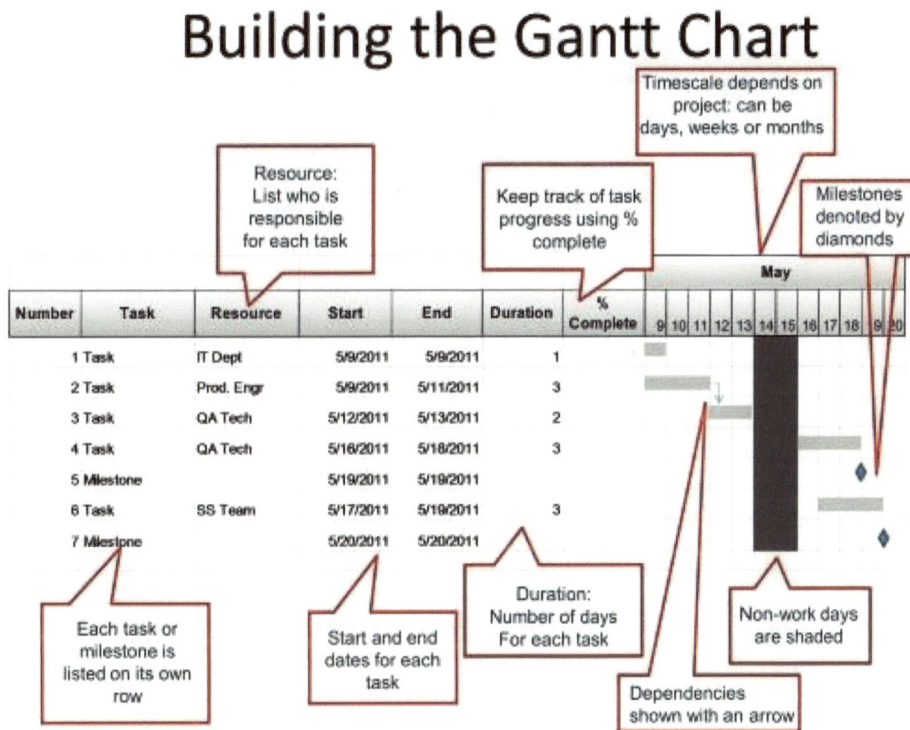

Figure 7.6 Gantt chart

7.3 The PERT Chart

For more complex tracking and planning, we can use a PERT chart, which stands for Program Evaluation and Review Technique. This project planning and scheduling technique was developed during WWII. The PERT chart shows the flow of steps or activities in a project, the time required to perform each step, the earliest start and end dates and the latest start and end dates for the activities. Using a PERT chart, a critical path can be identified, and progress can be measured against the project deadline date.

Activity Table

Activity	Predecessor	Time
A	--	10
B	--	5
C	A	15
D	B	10
E	D	5
F	C, E	5

Figure 7.7 Activity table

Often, we use an activity table as the input to the PERT chart construction, an example of which is shown in Figure 7.7. In this activity table, there are six activities, labelled generically A – F, which would be specific tasks for the project, such as develop a data collection plan, enter data into Excel, and so on. Some project steps can only occur after another step is completed. The predecessor column shows what steps must be complete before an activity can begin. For example, activity C can only begin after activity A is completed. Activity F can start only after steps C and E are finished. The time for each task is listed in the time or duration column. We can use the information in this table to create a PERT diagram.

In a PERT diagram or chart, each activity is placed in a node, as shown in Figure 7.8. The node has the activity name and the duration in hours or days, for example, at the top. Below this we have the earliest start and end dates, and the latest allowable start and end dates for each activity. Here, we are trying to hit a specific deadline, or specific project duration.

Key to PERT Node

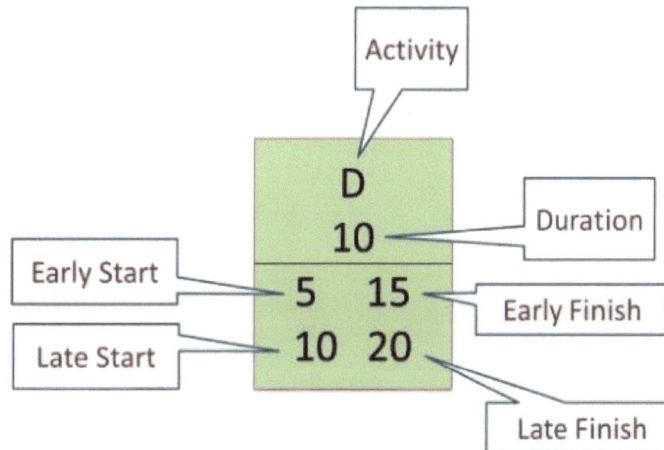

Figure 7.8 Key to the PERT node

Figure 7.9 shows a PERT chart with all the nodes filled out. (Wortman, 2004) Activity A has a duration of 10 time-units. We will call the units "days" for our example. The first line under the duration shows the earliest start and stop times. To create the chart, start by filling out the earliest start and stop time for each activity in this path. Activity A can begin at day zero, which is the project start. It takes 10 days, and the earliest it can be complete is at day 10. Next, we move to activity C. It is 15 days in duration. If Activity A ends at day 10, C can then begin on day 10. Add 15 days, and the earliest end time is 25.

PERT Chart Example

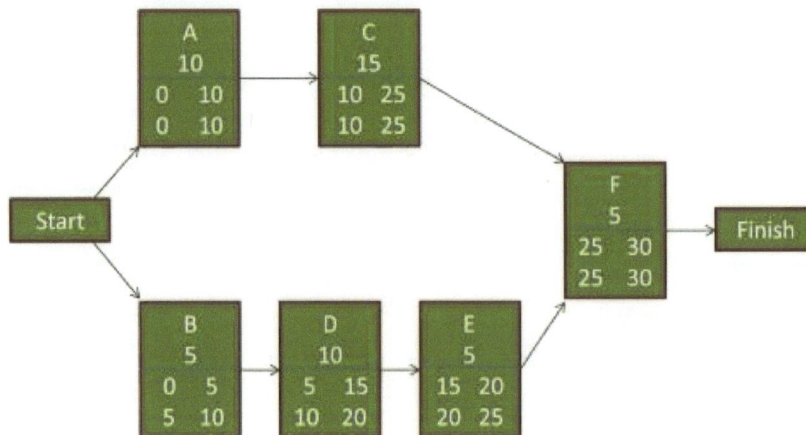

Figure 7.9 PERT chart example

Now onto the lower path: Activity B has a duration equal to 5 days, so we have 0 and 5 as start and end dates. Activity D has earliest start of 5 and earliest end at 5 + 10 = 15. Activity E can start at day 15 and end at day 20 since it only takes 5 days. Since C and E both must be done before F, as shown by the flow diagram, the earliest F can begin is day 25 since C will not be complete before then. Activity F takes 5 days, so the project can be completed is in 30 days. We set 30 days as our deadline.

The second line below the duration in each activity node shows the latest allowable start and end dates. The project must end on day 30, so that is the latest allowable end date. We subtract 5 days of duration for activity F and get latest allowable start to be 25 days. Along the A-C path we can see that the latest allowable end for C matches the latest start for Activity F. Subtract the 15-day duration for C, and we get 10. Moving to Activity A, the latest end can now be 10, and latest start is at 0. Note that the earliest and latest end and start dates are the same for activities A, C and F.

Now, looking at the bottom path, the latest end date for activity E is 25. If we subtract the duration of E= 5, we get the latest start equal to day 20. Activity D can thus end at day 20, subtract 10 days, and we get latest start at day 10, and activity B has latest end at 10 and latest start at day 5.

Using the information from the PERT nodes, we can now calculate *slack*, which is the difference between the late finish and early finish times. As shown in Figure 7.10, for Activity D, we can see that we have slack of 5 days, meaning that we have some wiggle room with the completion schedule for this activity. Using the extra 5 days here will not adversely affect the project completion.

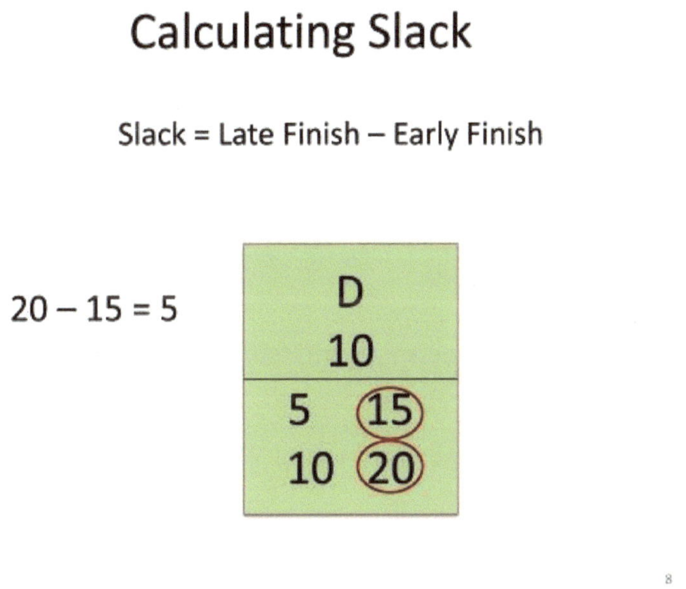

Calculating Slack

Slack = Late Finish – Early Finish

$$20 - 15 = 5$$

Figure 7.10 Calculating slack

We can continue to calculate the slack for each node: Note that the slack for each activity on a path is not additive: if Activity B uses up all 5 days of its allowable slack, there will be no slack left for Activities D and E.

The *critical path* is the path on the chart that has no slack. It must be completed on time: if the duration of any activity on the critical path slips, the whole project will be delayed.

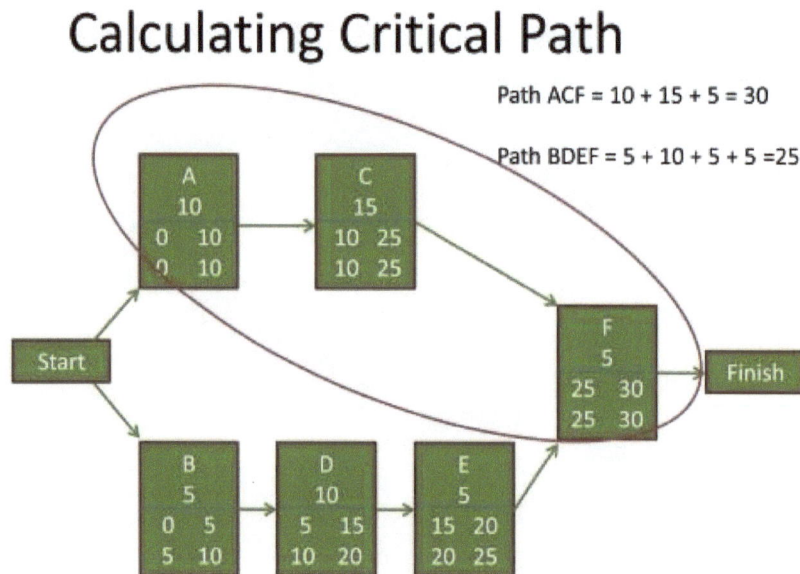

Calculating Critical Path

Path ACF = 10 + 15 + 5 = 30

Path BDEF = 5 + 10 + 5 + 5 = 25

A	C
10	15
0 10	10 25
0 10	10 25

F
5
25 30
25 30

Start

Finish

B	D	E
5	10	5
0 5	5 15	15 20
5 10	10 20	20 25

Note that slack = 0 for all activities on the critical path

10

Figure 7.11 Calculating the critical path

We can use the PERT chart to find the critical path one of two ways. In the first case, we list all the paths in the project. Here we have ACF and BDEF. Then we can calculate the duration of each path. The path with the longest duration is the critical path. As an alternative, we can calculate the slack for each path. The path with zero slack on each activity is the critical path. (Figure 7.11)

Chapter 7 Managing Projects © 2021 University Training Partners

Chapter 8 Mapping & Graphing Data

In this chapter, we will learn to transform physical information and numerical data into pictures that give insights into our processes. Process maps, swim lane charts, spaghetti diagrams, Pareto charts, run charts, histograms and scatter plots are a few examples of these graphical techniques. These tools span both the Measure and Analyze phases of a Six Sigma project.

8.1 The Process Map

Recall that the emphasis in Six Sigma projects is on process improvement. In the Measure step of a Six Sigma project, the team performs the vital step of mapping the process as it currently exists. This process map is more detailed than the quick process sketch used in a SIPOC diagram. Process maps should not be created by one person working alone in a cubicle or with a team shuttered in a conference room. Instead, a cross-functional team that understands different steps of the process is needed. Before the team starts writing process steps on sticky notes, it must get out of the conference room environment and walk the process.

We call this notion of walking the process *going to gemba*. Gemba is a Japanese term that means *the place where the work is done.* There is the process as it was designed, and there is the process as it is actually performed. Going to gemba allows the team to experience the true process steps firsthand — warts, workarounds and all.

Once the team returns to the conference room, it can write the process steps on the sticky notes and arrange them on the wall. The team can also use a few map symbols on the chart, such as the start and end point ovals, a rectangular activity box, or a decision diamond (Figure 8.1). The idea is to get the steps down without worrying too much about the mechanics of mapping.

Three Basic Map Symbols

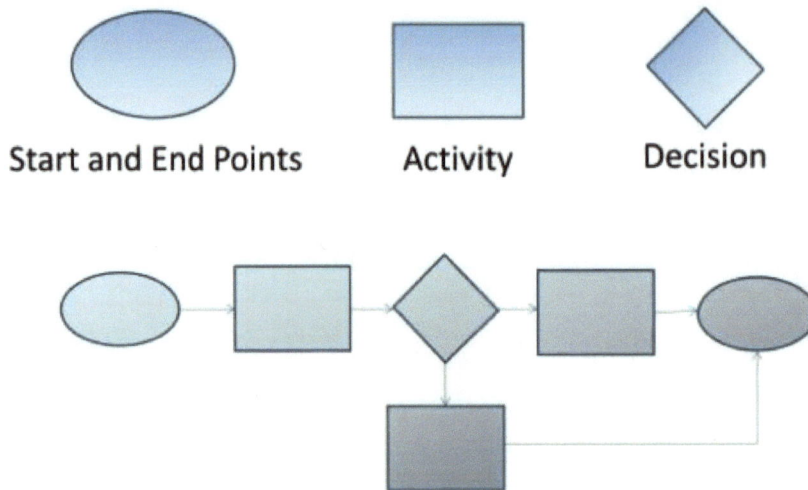

Start and End Points   Activity   Decision

Figure 8.1 Mapping symbols

When creating the process flow chart, make sure that the detail of the map to matches the team's purposes. Most likely, the process will need to be mapped in detail so that opportunities for improvement jump out at the team.

If there are quick fixes identified during the process mapping exercise, the team should go ahead and implement them. In these cases, there is no need to wait until the Improve phase. Quick fixes may include eliminating redundant steps, or steps that are unnecessary or detrimental to the process output. As for the suggestions and "Aha!" moments, make sure the team captures these in a parking lot – the ideas will come in handy once the team has reached the Improve step.

8.2 The Swim Lane Chart

The process map can be furthered refined by superimposing "swim lanes" that represent departments or functions. (Figure 8.2) The swim lanes help the team more readily see how the process steps and change of control flows across the organization. For example, are there too many iterative paperwork handoffs between departments? Can steps be combined? Can people work together instead of tossing their work product over the proverbial wall to the next department?

Swim Lane Chart

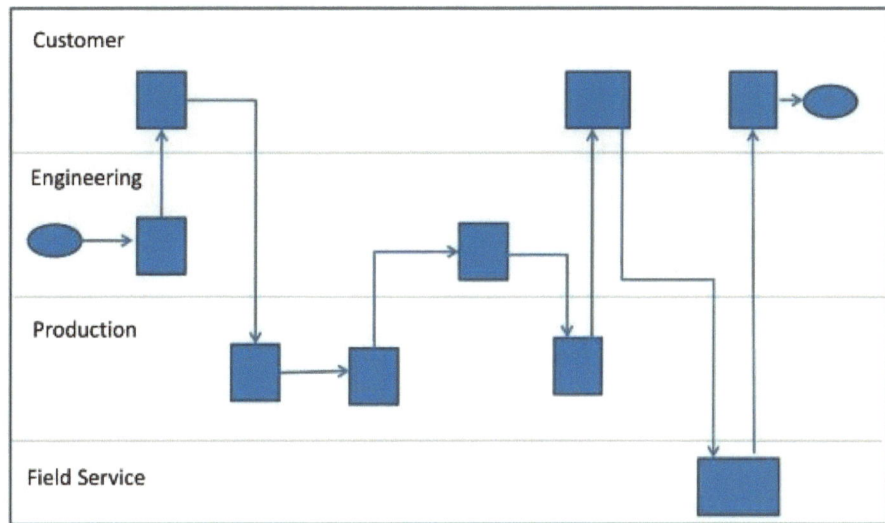

Figure 8.2 Swim lane chart

8.3 The Spaghetti Diagram

If our aim is to track the physical flow of a product through a process, we can use a spaghetti diagram (Figure 8.3). For example, using this tool, the team can track the movement of a patient through the emergency department of a hospital, the route a pick robot takes in a warehouse, or the movement of an assembly as it gets processed in the finishing department. The spaghetti chart can help the team identify poor layouts and excessive motion. For example, if the chart shows many crossing lines, it indicates that the physical layout is not optimal. Poor layout adds extra time and distance and may be dangerous if there are fork trucks involved. Having parts return to the same place repeatedly may indicate that the process steps should be rearranged or combined. The physical distance that a person or part travels through the process can also be measured.

Spaghetti Diagram

1. Indicate the location of each process step on a diagram of the workspace
2. Connect sequential steps with a line (the spaghetti)

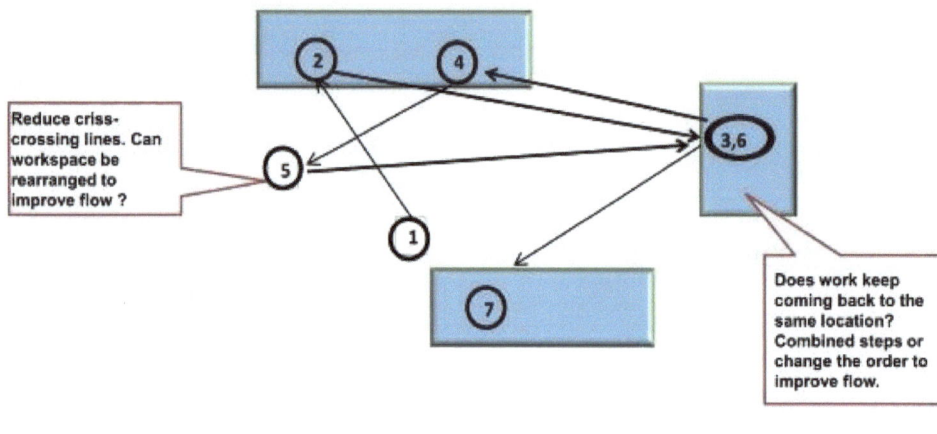

Figure 8.3 Spaghetti diagram

8.4 The Pareto Chart

A team can also use graphical analysis tools for the data it collects in the Measure phase. A simple check or tally sheet can be used to collect data on the frequency of certain events, as shown in Figure 8.4. Events might be defect types, rejection reasons, complaint types, and so on. The team can then turn the event frequency data into a Pareto chart.

Vilfredo Pareto was an economist active in the late 1800s and early 1900s who observed that 80% of the wealth in Europe was controlled by 20% of the population. In the 20th century, Joe Juran took this idea farther, realizing that this 80/20 rule also applied to quality problems. He called it the Pareto Principle, or *the vital few and the trivial many*. Using Juran's interpretation, eighty percent of the defects occur from 20% of the machines, or 80% of the complaints come from 20% of customers.

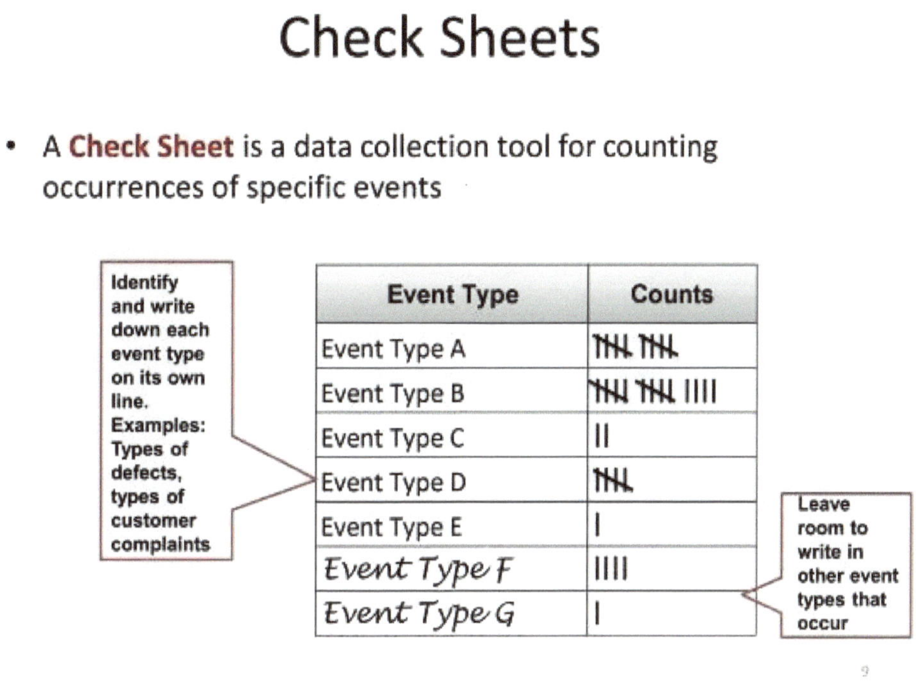

Check Sheets

- A **Check Sheet** is a data collection tool for counting occurrences of specific events

Figure 8.4 Check sheet

We can construct a Pareto chart based on tally sheets to help us concentrate on the biggest problems. The beauty of the Pareto principle is that the team does not have to eliminate every single cause or source of a problem to make a difference. Instead, it can make significant gains in quality by tackling the biggest contributors to the problem. The Pareto chart helps the team focus its efforts to address the main drivers of a problem.

Figure 8.5 shows an example of a Pareto chart from the Guest Services department at the Mirasol resort, based on the case study in Appendix A. The chart shows the number and type of customer complaints the hotel received in August. We can see that the main causes of complaints are *rooms not ready* and *AC problems*. These two sources account for 68% of all the complaints. What should Mirasol tackle first? The shuttle problem? Based on the chart, no.

Instead, the hotel can form two Six Sigma teams to address rooms not being ready and the air conditioning.

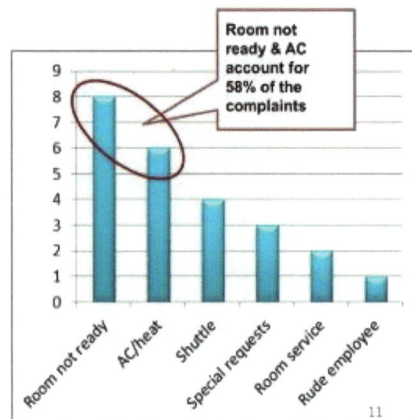

Figure 8.5 Pareto chart for complaints

A team can also create a cascade of Pareto charts if it needs to dig deeper into specific causes. For example, what types of AC complaints were there? That the AC was not cool enough or too cold? Was it too noisy? Was leaking water into the rooms? Using the specific complaints, Mirasol can create another Pareto chart prioritizing AC complaint types.

8.5 The Run Chart

A run chart is a quantitative variable plotted over time. Run charts form the basis of statistical process control charts and are helpful in detecting changes over time (Figure 8.6).

Of course, no one should not bring any charts to a meeting just because he or she created them. The charts presented should tell a story. It is a Black Belt's job to tease out business intelligence from data using graphs. For run charts, we look for patterns that give us insight into what is

Chapter 8 Mapping & Graphing Data © 2021 University Training Partners

going on. Typical patterns for run charts are shown in Figure 8.7. For example, a chart could show a trend over time, either increasing or decreasing, as shown in the first two plots in the top row.

Run Charts

- A **Run Chart** tracks how a measure or variable performs over time.
- A Run Chart has one variable which is always plotted against *TIME*

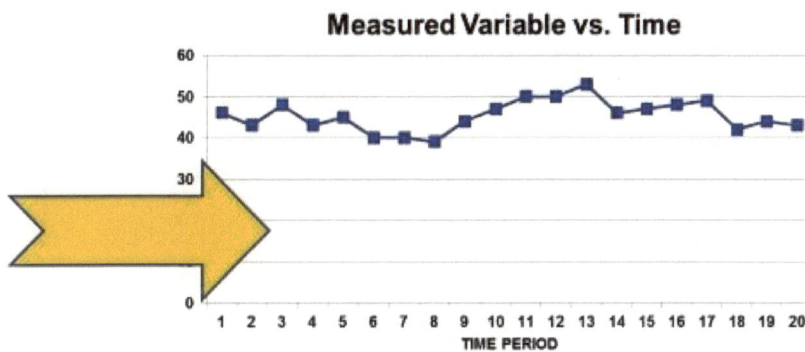

Figure 8.6 Run chart

Run Chart Patterns and Trends

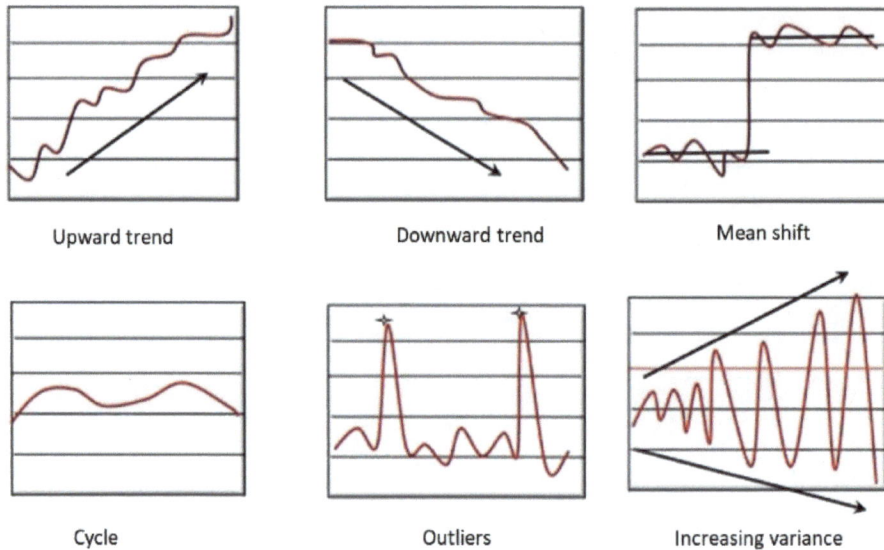

Figure 8.7 Run chart patterns and trends

The last chart in the first row of Figure 8.7 shows behavior that is called a *mean shift*, in which the variable is running at one level, and then suddenly it jumps either up or down. A mean shift indicates that there was a shock to the system that has caused a change. Mean shifts can be created by a change in supplier, a new machine, a new procedure, or a natural disaster, such as an earthquake in Japan, in Mirasol's case.

A run chart could also exhibit a cyclical as shown in the second row of Figure 8.7. Most industries are cyclical – certainly Mirasol is – along with education, banking, candy production, construction, and so on.

The second chart in the second row shows unusual points, called outliers. These are points that are much higher or lower than expected. There is often an assignable cause for the outlier, and it is the team's job to find it.

Chapter 8 Mapping & Graphing Data © 2021 University Training Partners

Finally, the last chart does not show a change in the mean, but rather, a change in the variability of the measurement over time. This behavior is certainly a red flag since it is the exact opposite of what we want to see in Six Sigma, in which the goal is to reduce variability and increase consistency.

Figure 8.8 shows a run chart for the number of guest complaints by month for Mirasol. What pattern do we see? There is a mean shift that occurs in June and remains in place well into the shoulder season. What happened in June to cause this? This will be the team's job to dig further into the data to find out.

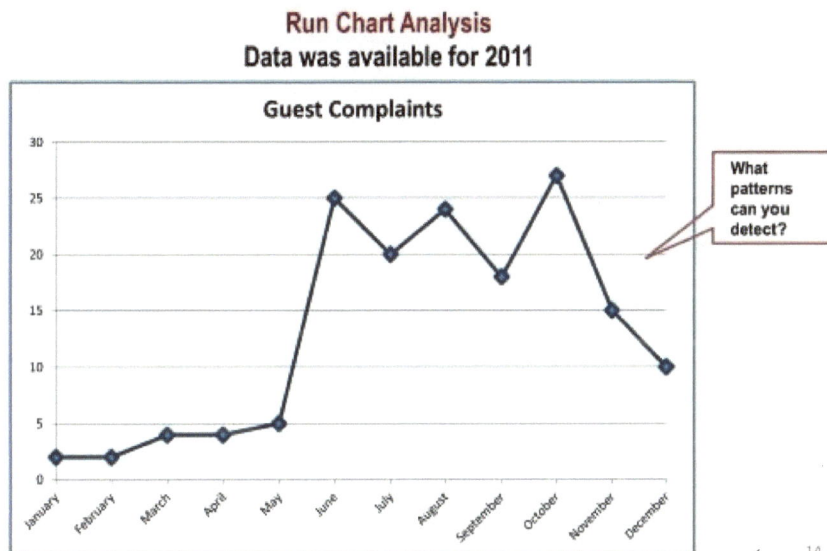

Figure 8.8 Guest complaints by month

Figure 8.9 shows percent occupancy by month, and we can see, as expected, that there is a cyclic pattern. To perform a complete analysis, the team should plot both complaints and occupancy on the same chart to see if the jump in complaints in June is due mostly to the higher occupancy rate. If not, the team would conclude that the jump in complaints is due to something other than simply more guests at the hotel. The two variables can be plotted on the same chart with different y-axis scales in Excel. This technique will be demonstrated in the course videos.

SIR Case Example
Occupancy by Month

Run Chart Analysis

Percent Occupancy by Month, 2011

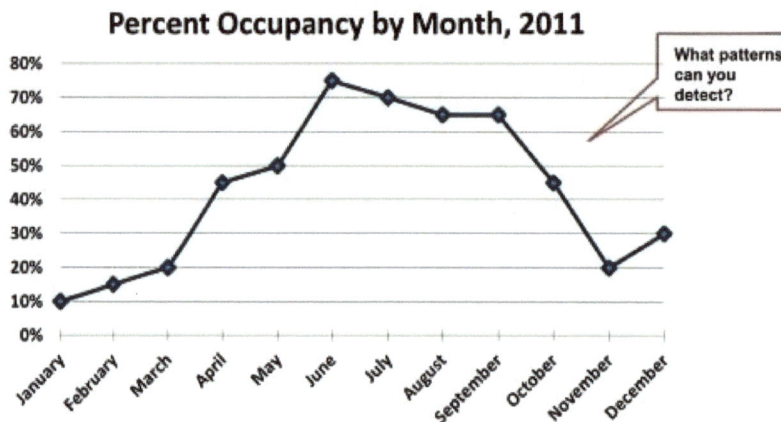

Figure 8.9 Occupancy by month

8.6 The Measles Chart

A measles chart shows a physical location of an occurrence of a defect or event. It is similar to a heat map, in which colors are used to show concentrations of various events or characteristics, such as crimes or land values. The measles chart is aligned with the spaghetti diagram in that both start with a physical layout of some type. The spaghetti diagram shows the flow through a system, whereas the measles chart shows the physical concentration of events such as defects.

The chart shown in Figure 8.10, is a rudimentary sketch of an automotive car hood. An assembly plant was experiencing quite a few paint defects on the hoods. To focus in on where

Chapter 8 Mapping & Graphing Data © 2021 University Training Partners

these defects were occurring, a measles chart was created by plotting the location of the paint defects found through inspection over several shifts. The type of paint defect was also tracked using symbols: an X, a triangle, and a circle. The resulting chart shows that most of the paint defects occurred in the left front of the hood, and most of those defects were paint runs.

Measles Chart

- A **Measles Chart** shows where defects are physically occurring

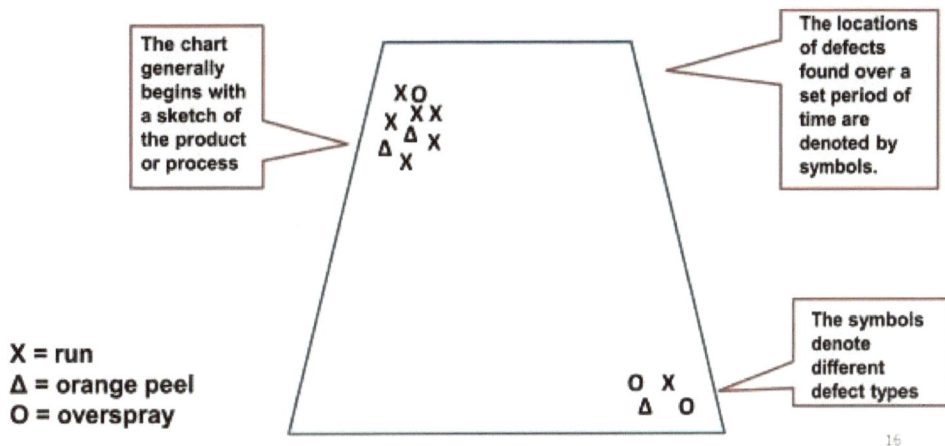

The chart generally begins with a sketch of the product or process

The locations of defects found over a set period of time are denoted by symbols.

The symbols denote different defect types

X = run
Δ = orange peel
O = overspray

Figure 8.10 Measles chart for paint defects

Now the team assigned to address the problem can concentrate its efforts on the paint robot that sprays the front corner of the hood to find the root cause.

Figure 8.11 shows a measles chart constructed by the "Reduce AC Complaints" team at the Mirasol resort. Most of the AC complaints originated from rooms along one hallway on the second floor of the hotel. Now the team can begin to check if those rooms are all serviced by the same compressor, or if the ducting to those rooms is blocked, or if those rooms happen to face the south.

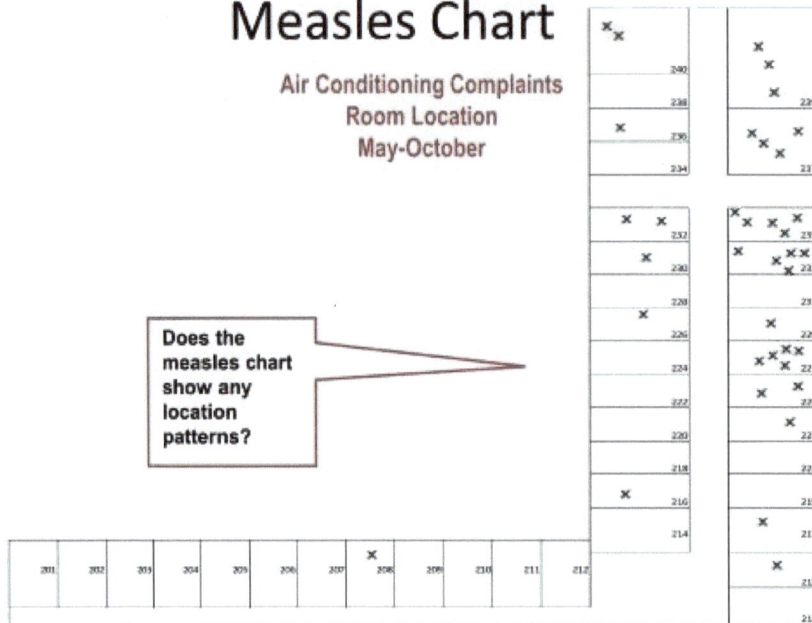

Figure 8.11 Measles chart for AC complaints

8.7 The Histogram

A *histogram* is a plot of a continuous variable that shows the shape, or distribution, of the data. A histogram is generated from a frequency table in which the continuous data is categorized into different classes, or bins, and then the number of data points falling into each category is counted. The x axis of the histogram lists the classes in numerical order, and the y axis shows the frequency of data points in each class.

Histograms

- Graphical display of a frequency table
- Give a sense of the shape of the data distribution

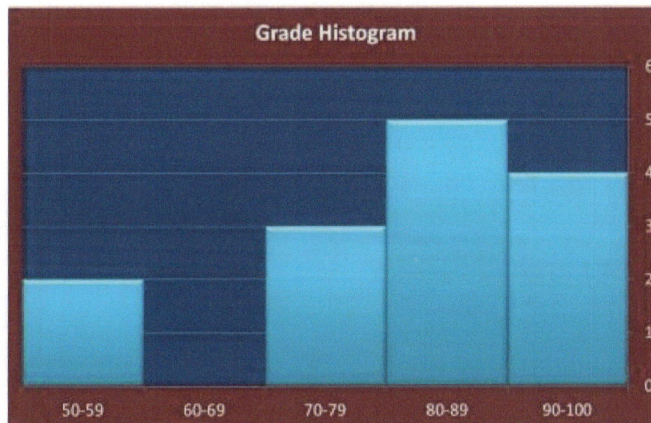

Figure 8.12 Histogram

The histogram in Figure 8.12 shows the distribution of class grades after a midterm exam. It is readily seen that the most frequent score was in the 80-89, or B, range, and that there were two students who scored in the 50-59 range. In addition, the distribution is not symmetric, but rather, skewed to the left, or low end.

In general, the histogram gives us a sense of the central tendency of a distribution, the variability of the data (based on how wide the histogram is), and whether the distribution is symmetric or skewed. The skew is determined by the direction of the tail of the distribution, as shown in Figure 8.13.

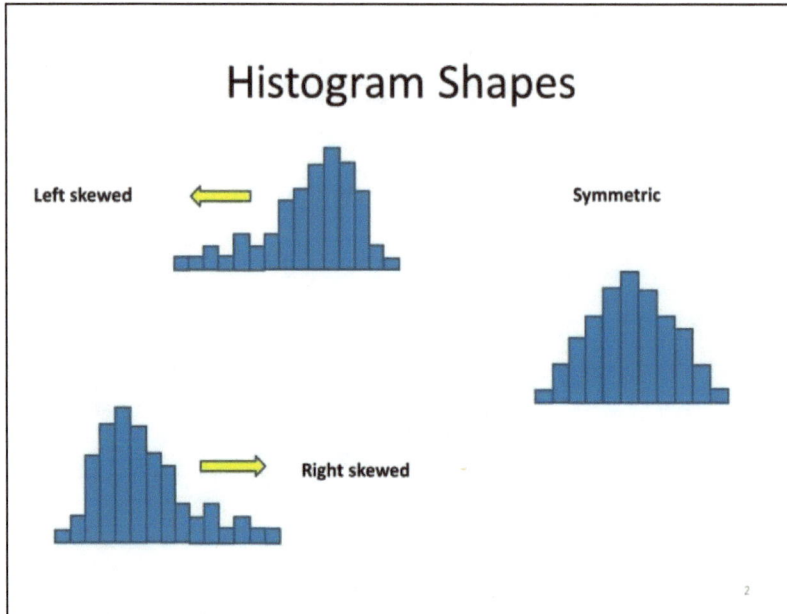

Figure 8.13 Skew of distributions based on histogram shape

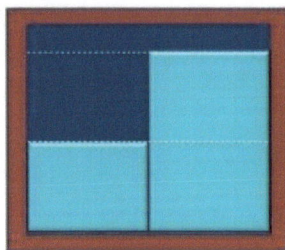
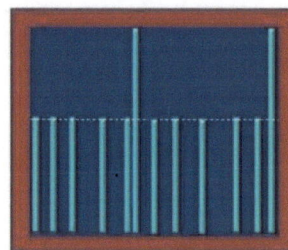

- Aim for k = 5 to 20 categories

- Can use a rule of thumb: $k \cong \sqrt{n}$

Figure 8.14 Choosing the number of categories for a histogram

When constructing a histogram, make sure to categorize the data into several bins so that the underlying distribution of the data can be visualized. If you choose two few bins, the histogram will look lumpy. Conversely, if you choose too few, the histogram resembles the teeth of a comb (Figure 8.14).

There is a rule of thumb about choosing the number of categories, and that is to take the square root of the number of data points. This figure will give us a good starting point for the number of categories to use. Of course, using a statistical package like Minitab will eliminate this step, since the program chooses the number of categories and their endpoints automatically, although the histogram can be customized to suit our needs, as we will learn in the Statistics for Black Belts course.

8.8 The Scatter Plot

If we have two quantitative variables that are collected in pairs, we can plot the data on the x-y grid to create a scatter plot. For example, if height and weight data is collected from a group of people, each person in the group would have two data points associated with him or her, and this x-y pair would be plotted as a point on the graph. Home price and square footage, or room occupancy and revenue, are also examples of x-y pairs. Once the data is plotted in the x-y grid, form of the relationship between the two variables can be described. In other words, the scatter plot might suggest a straight-line relationship or a curvilinear relationship between the x and y variables.

If the points suggest a straight line (also called a first-order relationship), we can then go further and make statements about the direction and strength of the linear correlation. Plots of height and weight, home price and square footage, and room occupancy and revenue tend to follow a straight line. In each of these examples would exhibit a positive linear correlation, since the two variables tend to move in the same direction: as one goes down, the other tends to decrease, or as one goes up, the other tends to increase.

A plot of monthly electricity and natural gas costs, however, would display a negative correlation since these variables move in opposite directions. In the summer when the air conditioning is running, electricity costs will increase, while the gas heating bill would decrease. Conversely, the gas heating bill will increase in the winter months while electricity costs will go down. The classification of correlations is covered in more depth in the online portion of the class.

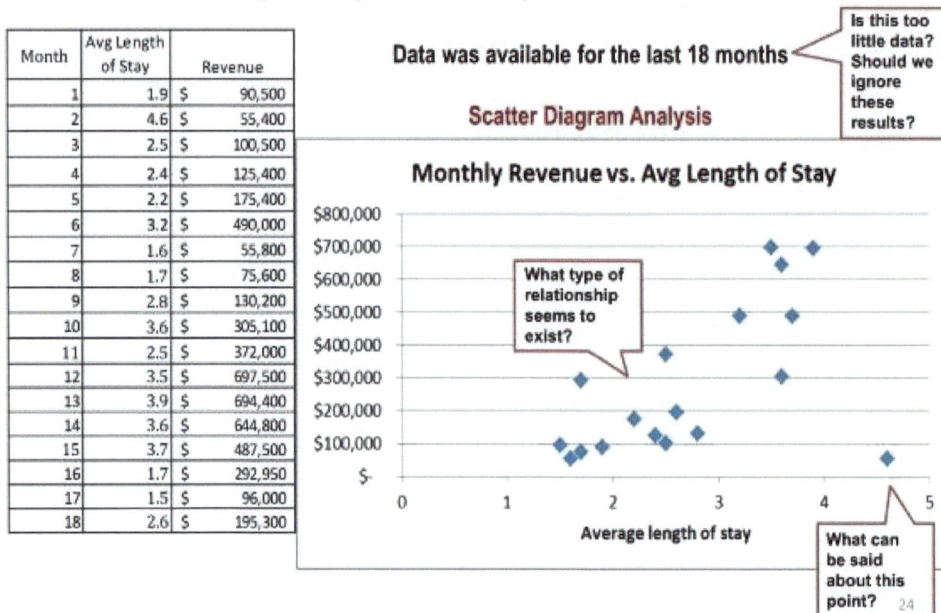

Figure 8.15 Scatter plot for Mirasol

Figure 8.15 shows a scatter plot created from the monthly average length of stay and total revenue at the Mirasol resort. Average length of stay is plotted on the x axis and revenue on the y axis. The plotted points suggest a straight-line relationship, as well as a positive correlation. There is a fair amount of scatter, so we might characterize this as a weak positive correlation. There is also an outlier, or a point that does not follow the pattern of the others. The team could go back and investigate this point to see if it were due to an assignable cause. An analyst cannot simply delete points because they don't follow the rest; however, if there is a data entry error, or a special cause, then there would be justification for removing the point.

The main purpose of a scatter plot and a regression equation is not to report on the data we have collected, but rather, to calculate a mathematical model that allows us to predict y given an x *in the future*. Using Excel, we can create a scatter plot, superimpose a fitted line over the points, calculate a sample linear correlation coefficient, and a fit a regression equation. These procedures will be demonstrated in the online course.

There are a few cautions that must be recognized when interpreting correlations and scatter plots. First, a correlation is a measure of the linear relationship between two variables. If a scatter plot indicates that a curved relationship is present, then calculating a linear correlation is not appropriate. Second, when interpreting a linear correlation, we should be careful to not confuse correlation with causation. Just because x and y are related does not mean that a change in the x variable *causes* the change in the y variable.

For example, consider the relationship between ice cream sales and shark attacks shown in Figure 8.16. When plotted, these two variables exhibit a strong positive correlation. That does not mean that the number of shark attacks can be reduced by decreasing the sales of ice cream. In actuality, the two variables are both related to a third unrecorded variable, called a *latent variable*, which is, in this case, is season of the year. It is through their mutual correlation with season that shark attacks and ice cream sales are correlated.

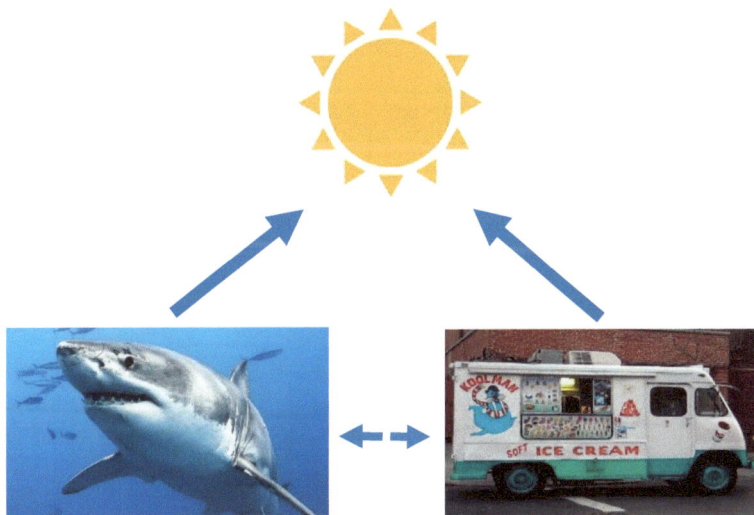

Figure 8.16 Latent variable diagram

We also should take care not to extrapolate by predicting a y from a value of x that does not fall in the range of the original data. Just because the relationship shows a positive correlation, for example, does not mean the relationship will hold for all values of the x variable. For example, using the scatter plot in Figure 8.18, we cannot reliably predict the revenue for an average stay of 14 days at the Mirasol resort, since there is no information about this length of stay in the data. We do not know the form of the x-y relationship in the region of a 14-day length of stay.

Chapter 9 Finding Root Causes

Collecting and analyzing data helps the team narrow its focus in solving the problem at hand. As shown in Chapter 8, graphical techniques such as measles charts help the team pinpoint where defects are occurring, and run charts give insight into the timing of defects. In addition, scatter plots may show the team an unexpected relationship between two variables.

These graphical techniques show the team the *symptoms* of a problem (Okes 2004). A symptom is the effect of a system or physical failure and may include defects, customer complaints, returns, or warranty claims. In the Analyze phase of a project, the team must identify either physical or systemic causes of the symptoms. Systemic causes involve the organization's policies and process designs.

There are several methods for generating and sifting through root causes of the problem, including brainstorming, fishbone diagrams, flow charts, why-why diagrams, is-is not diagrams, and root cause analysis diagrams. Whatever method is used, the team must keep in mind the principle of Occam's razor, in which the simplest explanation is usually the most likely, as exemplified in Figure 9.1.

"If we hear hoof beats off in the distance, it is likely to be a horse, not a zebra."

Figure 9.1 Occam's Razor

9.1 Brainstorming

The goal of brainstorming is to generate as many ideas as possible in the time allotted. Brainstorming sessions are ubiquitous, but it has been shown that these sessions are often not very effective.

There might be a dominate personality who takes over the session, or an intimidating environment in which employees are reluctant to say what they really think. In addition, there are often naysayers who shoot down ideas in the brainstorming session, thus impeding creativity. Sometimes ideas do not get written down on the board, or the group loses its momentum by discussing the merits of each idea.

Brainstorming approaches should be matched to the group dynamics. If the team is a small, homogeneous group, then a freewheeling approach to generating ideas might be productive. If the group is larger, and/or made up of some introverts and extroverts, a round robin approach in which members take turns proffering ideas may be better. Having team members write down ideas individually is the preferred approach if some team members feel intimidated by others, or if the subject itself is sensitive or fraught with office politics. This way, everyone's ideas can be captured in an anonymous way.

As a Black Belt, team leader and facilitator, you will have to gage the group dynamics to see which brainstorming approach will work best.

9.2 De Bono's Methods

Edward de Bono has written and lectured extensively on creativity and thinking skills for teams. Two of his methods, random stimulation and Six Thinking Hats are discussed here.

The **random stimulation method** forces a team to deviate from their usual thinking processes. A random noun is chosen from a dictionary, and the team is asked how the problem or issue at hand relates to that word. For example, if the team at the Mirasol resort is trying to brainstorm root causes of the "rooms not ready" problem, it may be asked to find ways to associate the problem with the randomly chosen word "speech." Team members may brainstorm associations such as "communication," "language barriers," "management not doing what it says it will," and so on. The idea of the exercise is to force team members to find patterns in novel ways, thus stimulating creativity of thought.

De Bono's second technique is called **Six Thinking Hats**, illustrated in Figure 9.2. Each hat represents a unique way of thinking about a problem. The entire team will "wear" a hat and brainstorm solutions using the hat's perspective. Then the team will move on to the next hat, until all six are covered. By wearing all six hats, the team is assured that is has thought about the problem from all sides. (McCarty, 2005).

To perform the Six Thinking Hats exercise, the team starts with the procedures Blue Hat to lay out the groundwork for the thinking exercise. For example, in what order, will the hats be worn? How will the ideas be captured? By whom? Then the team will move on to the other five hats.

For example, the team at Mirasol that is addressing the "rooms not ready at check in" problem may choose to wear the data-oriented White Hat next. At this point, the team would think about the causes of the "rooms not being ready" problem using customer complaint history and room turnover times. Then, wearing the creativity Green Hat, the team would switch to creative solutions, such as installing new tracking technology or self-vacuuming rugs. The team would continue its session until all the other hats were worn. The team would end the session with the Blue Hat, summarizing what it has learned.

9.3 Mind Map

A **mind map** is a graphic representation of ideas and how they relate to each other. This map displays pictures, symbols and words interconnected with lines. It is meant to recreate the way our brains store and retrieve information using a network of associations and pictures. The consultant Tony Buzan and his company are promoters of this type of brainstorming tool.

A single user can create a mind map using paper and several colored pens. A team can produce a map using a large white board. The central idea or problem is placed in the middle of the paper or white board. Then, lines radiate out from this central point as team members think of topics associated with the main idea. These branches are labelled. Ideas from this branch can be further divided into more branches. Words, pictures, or symbols can be used as well as different fonts and colors. Once the draft map is created, the team can reorganize and edit to arrive at the final mind map.

Six Thinking Hats	Thinking Perspective
Blue Hat 	Concentrate on organization, procedures, and control (begin & end with this hat)
White Hat 	Use only data and information at hand (no opinions)
Yellow Hat 	View ideas with optimism and positivity, concentrate on the benefits of proposed solutions
Black Hat 	Exercise critical judgment, play devil's advocate
Red Hat 	Express feelings or voice intuition
Green Hat 	Exercise creativity, concentrate on innovative, out-of-the-box solutions

Figure 9.2 DeBono's Six Thinking Hats

Figure 9.3 shows an example of a mind map created to identify the barriers to achieving a Lean Six Sigma Black Belt certification. Colors, pictures, and words are used to capture the ideas of the group.

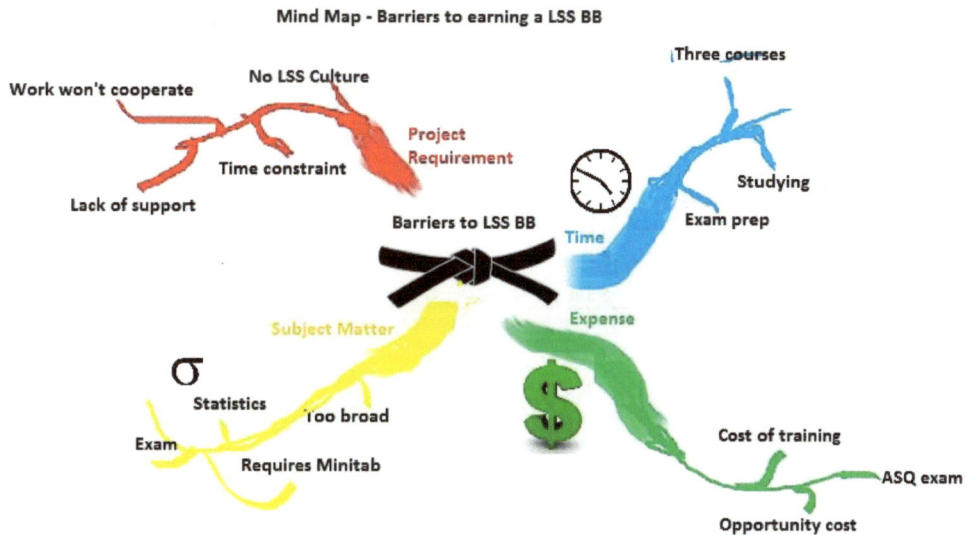

Figure 9.3 Mind map

9.4 Fishbone Diagram

One of the most popular techniques for finding root causes is the **fishbone diagram**. The fishbone diagram, also called the cause-and-effect diagram or Ishikawa diagram, requires the team to brainstorm ideas and place them in categories.

A sketch of the fishbone diagram is drawn on the whiteboard, with the problem or issue written at the head of the fish. The four major bones and the tail can be categorized according to one of several schemes. The standard labeling method uses 5M's: Manpower, Methods, Material, Machinery and Mother Nature (also known as Environment), as shown in Figure 9.4. Alternate schemes can be used. For example, we can use People, Policies, Procedures, and Processes.

The team brainstorms possible causes of the problem or issue, and then puts each idea on the appropriate bone. Ideas that are subsets of the original are drawn with a branch from the original bone. After all ideas are exhausted, the team can go back and addresses all the causes listed and ask "Why?" to drill down to a root cause.

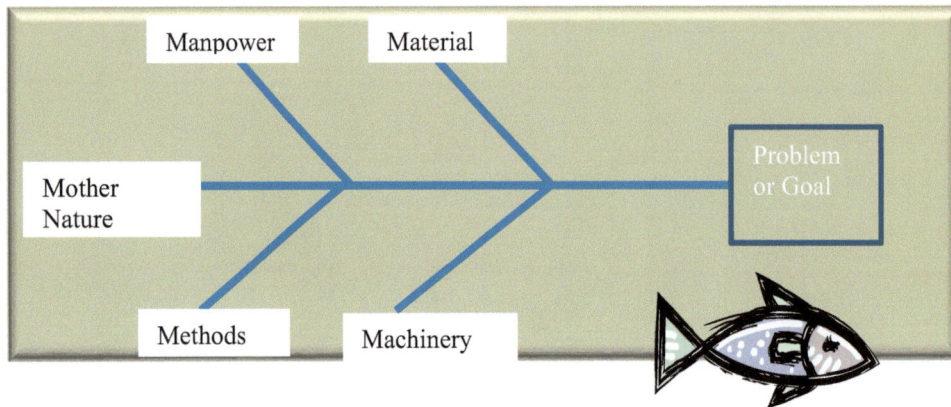

Figure 9.4 Fishbone diagram

9.5 Why-Why Diagram

The team can also use a **Why-Why diagram** that documents the response to the "5 Why" exercise. This technique allows the team to go through the root cause thought process incrementally, instead of leaping to conclusions.

Using the following procedure to construct a Why-Why diagram:

1. Write the problem or issue on the left-hand side of the white board.

2. Ask "Why?" List all causes on separate sticky notes and place in a column to the right of the problem.

3. Now, treat each of these causes as new problem statements. Ask why and place these causes to the right of the other column. Draw arrows connecting problems and causes.

4. Continue the diagram until a physical cause of the problem is found.

5. The system cause can then be found, if needed, by going a few steps further. Stop when the team reaches a fundamental, or system cause, such as "company policy."

The Why-Why diagram technique can be combined with a process map. The process map shows the flow of the process over time, whereas the Why-Why diagram shows a structural relationship. (Okes, 2009)

9.6 Is-Is Not Matrix

The **Is-Is Not Matrix** helps the team identify who, what, when, where and how a problem is occurring.

To use the Is-Is Not matrix, the facilitator describes the problem or event in a few words on the board. Then, the team is led to fill out the chart noting what is true about the situation, and what is not true. The completed matrix can give valuable insights into the nature of the problem. A template for the matrix is shown in Figure 9.5.

Statement or description of the problem goes here	IS What does occur?	IS NOT What does not occur?	Distinctions Is there anything that stands out as odd?
What is affected?			
Where does the problem occur?			
When does the problem occur?			
What is the extent of the problem?			
Who is involved?			

Figure 9.5 Is-Is Not template

9.7 TRIZ

The term **TRIZ** (pronounced "trees") is an acronym in Russian that translates to the Theory of Inventive Problem Solving. It is a very through and involved system for producing solutions to difficult problems or developing novel ideas. The technique was developed by Genrich Altshuller in the 1940s. Altshuller researched millions of successful inventions and innovations and developed a list of basic principles that successful inventions possess, and a list of fundamental problems that organizations encounter. For example, TRIZ identifies 39 product parameters that can be in technical conflict, and 40 principles that can be applied to solve these technical conflicts. There are also 76 inventive standards. See Cameron (2010) for an interesting and detailed introduction to the topic.

9.8 Interrelationship digraphs

The **interrelationship digraph**, also known as a **relations diagram**, shows the connections between ideas. Brainstormed ideas concerning an issue or problem are used as the input to the digraph. The ideas can be created from scratch, or the inputs could be gathered from other tools, such as items on the smallest bones of a fishbone diagram, or the ideas listed in an affinity diagram.

The ideas are placed on separate sticky notes, and one at a time, cluster similar ideas are clustered together on a white board, leaving space between notes to draw lines and arrows.

Then, the ideas are connected. For each idea, an arrow is drawn to other ideas that are caused or influenced by the idea at hand. At the end, there will be several incoming and outgoing arrows for each idea. Count the number of incoming arrows and outgoing arrows and indicate the totals below each sticky note as # incoming/ # outgoing.

The ideas that have the most outgoing arrows are thought of as *basic causes*. The ones with most incoming arrows are *effects*.

Figure 9.6 shows an interrelationship digraph that was created by the team at the Mirasol resort to address the air conditioning complaints problem. It appears that "lack of maintenance" is a major driver, or cause, of problems since it has the most outgoing arrows. The note "thermostat not turned down" is a key effect with the most arrows leading into it.

Reasons for AC Complaints

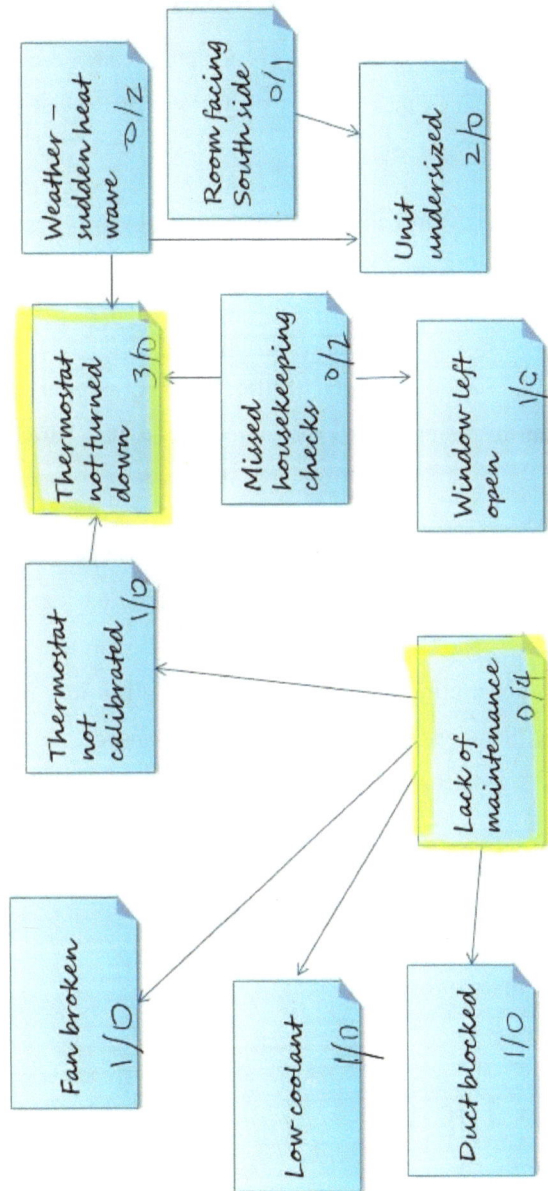

Figure 9.6 Interrelationship digraph for AC

9.9 Deciding Among Possible Causes

Once a list of potential causes has been created, the team needs to address each item to determine root causes. Many times, there is not just one cause, but several that contribute to a problem.

The team can systematically investigate the potential causes using selection tools. For example, the team can use a Pareto diagram to prioritize the causes based on dollar impact, likelihood, or severity, for example.

The team can also choose to use a **prioritization matrix**, introduced in section 3.3. Using the matrix, the team can determine which causes to address first based on selection criteria and weightings.

Last, a **payoff matrix** can be used to make decisions on what actions to take. The payoff matrix has its roots in game theory. It is a 2 x 2 table with effort on the x axis and payoff on the y axis, as shown in Figure 9.7. The team categorizes each solution in terms of Effort and Payoff. If a solution requires minimal effort and has a high payoff, then it is in the green zone, and is a number one priority. Solutions in the low effort and low payoff, or high effort and high payoff are prioritized next. Solutions falling in the high effort low payoff quadrant are not considered.

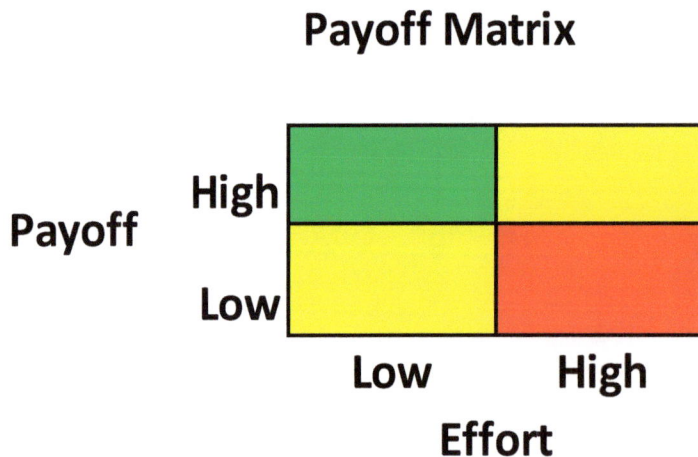

Figure 9.7 Payoff matrix

Chapter 10 Implementing Lean

It makes sense to incorporate Lean principles and tools into Six Sigma projects. Lean tools can help streamline and mistake-proof processes and are a great companion to the Six Sigma tools introduced thus far in the text.

10.1 Lean Principles

There are five principles of Lean, described as action items below.

1. Specify value, in which value is perceived from the end user's point of view.

2. Identify the value stream, which includes all the activities performed from the time of customer order to delivery. Every task performed can be classified as value-added or non-value added.

3. Create a flow to let products – parts or paperwork or information – move through the system without stopping.

4. Pull from the customer, meaning that products are made-to-order and shipped and not placed in inventory. In this sense, the customer could be the end user, or the next machine down the line. Nothing will be produced until the customer needs it.

5. Seek 100% quality and perfection in operations. Contrast this to the Six Sigma goal of 3.4 dpmo, after the shift. The Lean standard is much more difficult to achieve. The standard incorporates the idea of continuous improvement in which we implement improvements each day to incrementally achieve perfection.

Non-value-added activities or policies are classified as waste. In Lean there are eight different sources of waste that the team can identify and then eliminate.

An "elevator speech" for Lean would include the fact that Lean is process focused, team focused, and stresses hands-on manipulation of the system, versus, say, data analysis and statistics. There is also an emphasis on continuous improvement.

Frequently used Lean tools include Eight Wastes, visual management, value stream mapping, 5S and kaizen events, mistake proofing, and quick changeover, also known as SMED (for single minute exchange of dies.)

10.2 Eight Wastes

As we have noted, the emphasis of Lean is on reducing waste, and there are eight named wastes: transportation, inventory, motion, waiting, overproduction, over-processing, defects, and skills (Figure 10.1). These wastes do not add value to the product or service and should be eliminated. The first letters in each waste spell the name TIM WOODS which is a useful mnemonic. We will now define each of the wastes.

Eight Wastes

- Focus on eliminating non-value added activities

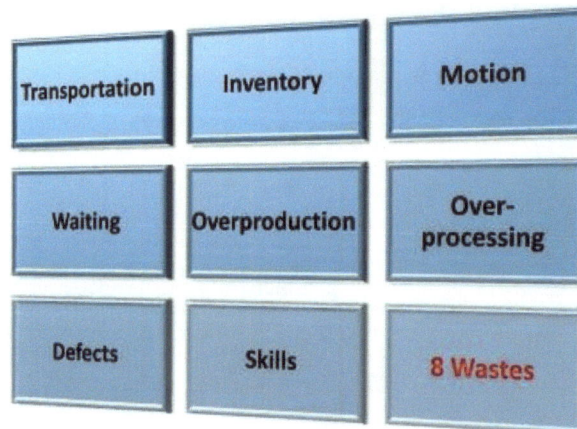

Transportation	Inventory	Motion
Waiting	Overproduction	Over-processing
Defects	Skills	8 Wastes

21

Figure 10.1 Eight wastes

Transportation waste is the movement of things, whether they are physical materials, or paperwork, or electronic information. Moving things from place to place does not add value to the product or service. In fact, there is a higher probability of the material getting lost or damaged the more it is moved. There might also be delays in availability because objects are in route. This may lead to another waste, waiting. One of the underlying reasons for transportation waste is a poor layout.

The next waste is *inventory*. This is an excess of things, whether it is parts, supplies, equipment, paperwork, or data. Inventory costs money to accumulate and to store, so a major effect of this waste is reduced cash flow. Inventory waste can result in lost production as workers look for items in a crowded storage area. Inventory may also get damaged or become obsolete before it can be used. A tell-tale sign of excess inventory is stockpiles of materials, or disorganized storage areas. The root cause of this type of waste is a just-in-case, ace-in-the-hole type of mentality, which might also be driven by an unreliable supply chain.

The next waste is *motion*, which is the movement of people, (in contrast to transportation which is the movement of things). Often, these two wastes occur together, especially when people, not machines, are moving the materials around. Motion waste could also stem from a poorly designed process, in which the operators must use excessive motion to get their jobs done. This is due to poor process layout and can result in worker injury.

Waiting waste is having people waiting for other people. This waste can occur when meetings start late, or when people must wait for information before they can move on to the next step, or when people wait for products or machines to be available. We can also think of this waste another way, in which information, products or machines are waiting for the people to act upon them. Waiting increases cycle times and may increase overtime hours. This waste can be due to unbalanced workloads, or a push environment, in which products are sent downstream even if the next process is not ready for them.

Next, we have *overproduction* waste. A direct consequence of overproducing is inventory waste. It can also trigger waiting waste for as work-in-process parts wait to be finished.

Over-processing waste is doing more than the customer is willing to pay for. This includes performing inspections. The customer is not willing to pay more for inspection; they expect the product to be made right in the first place. Over-processing waste could also be incurred by adding features to a product or service that are not valued by the customer. Adding unnecessary features could occur because customer requirements are unclear, or because there is a lack of VOC data. Over-processing results in longer lead times for delivery and a frustrated workforce being asked to perform tasks that are not adding value.

Of course, a major contributor of waste is *defects*. When we have defects, we increase internal and external failure costs, and have dissatisfied customers.

The final waste is the waste of *skills*. This waste occurs when employees' skills and aptitudes are not used to their fullest extent. This waste can result in frustrated workers, absenteeism, and turnover.

As the team analyzes a process or department for waste, it will see that there are many interconnections among the various forms of waste. When there is transportation waste, chances are there is also motion waste. If these two wastes exist, there might also be waiting waste. Overproduction leads to inventory, and so on. Many of the root causes of these wastes also overlap, which is a positive thing: by addressing a single root cause, the team may be able to reduce or eliminate several forms of waste at once. For example, lack of training is a root cause for transportation, motion, waiting, defects, and skills wastes. Root causes for each of the eight wastes are summarized in Figure 10.2.

Waste	Root Cause
Transportation	• Poor layout • Lack of cross-training
Inventory	• Just-in-case mentality • Unreliable supply chain
Motion	• Poor layout • Lack of cross-training • Insufficient equipment
Waiting	• Too many handoffs • Push environment • Unbalanced workloads • Lack of cross-training
Overproduction	• Lack of systems thinking • Push environment • Individuals valued over teams
Over-processing	• Lack of trust • Unclear customer requirements
Defects	• Poor training • Non-standard work • Lack of job aids • Poor communication
Skills	• Lack of trust • Lack of training • Silo thinking

Figure 10.2 Root causes of waste

10.3 Visual Management

A major tenet of Lean is visual management. Visual management makes the status of inputs, outputs, or the process readily apparent at a glance. In general, visual management techniques are inexpensive, simple, unambiguous, and immediate. By using visual management, problems are easily detected so they can be corrected quickly.

Visual management can be used in the factory certainly, but it also can be used in an office, a hospital, a restaurant, or any workplace.

The first feature of an organization that uses visual management is standardization, in which the ways things are done and what signals mean are consistent. There also is a safe environment because things are clean and organized. Progress and performance are easily known. In a visual factory, anyone looking out over the factory floor can notice immediately which machines are running, which are waiting for parts, and which are shut down. Next, in a visual environment, there is no blame assigned for finding defects. To the contrary, workers are praised for finding problems and correcting them or calling for help.

In a visual environment, action occurs at a signal. If a red light goes on at a machine, engineers and maintenance personnel come immediately to find out what the problem is. For example, consider the call light on a check-out counter at a grocery store when a cashier needs a price check. Someone immediately comes to take the item and find its price.

There are many tools used for visual management, often low-tech ones, like using colored tape on the factory floor to show where inventory belongs, to show where someone shouldn't stand, or where the fork trucks will be running. Racks and bins can be color-coded by part type. Indicator lights, known in Japanese as *andon*, are on every machine to signal status, as well as operator call lights. In addition, there are pictographs of work instructions placed at every workstation.

Maintenance charts show the latest service performed and statistical process control charts show the current process performance. There are also kanban bins that indicate whether parts are needed, and display boards showing current production levels, quality levels, process time, and so on.

10.4 Value Stream Mapping

The next Lean tool is value stream mapping, which is similar to a process map, but with much more information. The value stream map captures process steps, but it also shows how information, work, and materials flow through the process. These maps are used to identify and quantify waste. When creating a value stream map, the team records the process as it exists on a present state map to find opportunities for improvement projects. After analyzing the current state, the team draws a future state map, the what-should-be map, to use as a blueprint for improvement.

To create a value stream map, we choose one product, service, or product family, draw the process flow, then add in material flow, information flow and process data such as yields, cycle time and lead times. From this detailed map, we will be able to see waste in the system (Nash, 2008).

Next, we create a future state map showing the ideal process. We put Kaizen Bursts on this map to remind us of what needs to happen for the future state to become a reality. This map is the team's plan for the next six to twelve months.

There are quite a few symbols used on a value stream map, including inventory, truck shipment, information flow, and kaizen bursts, for example, as shown in Figure 10.3.

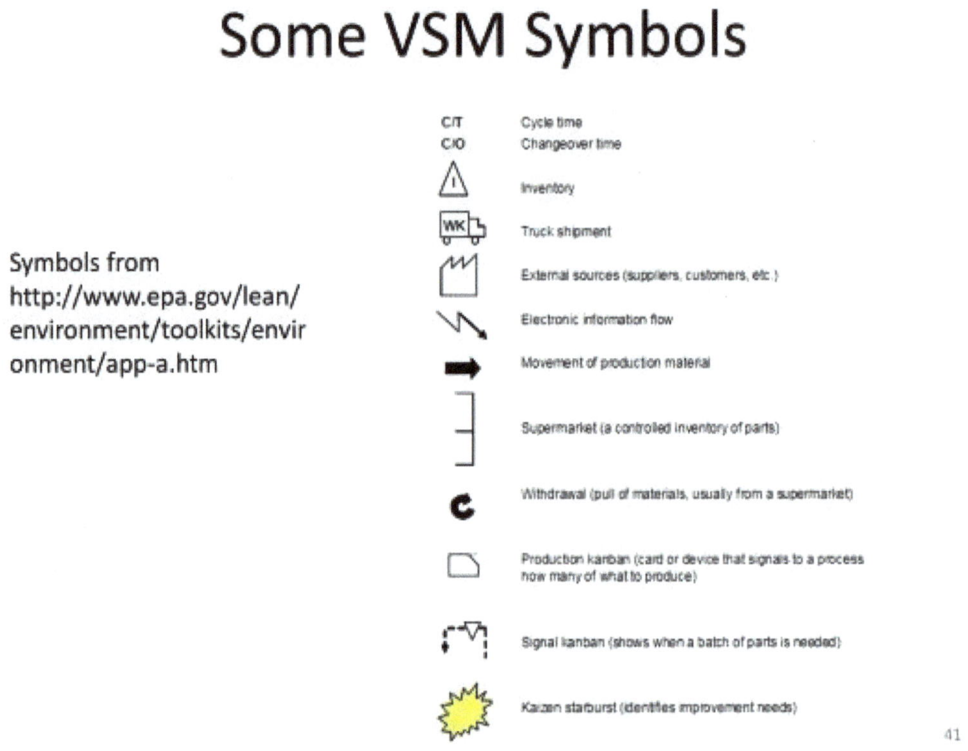

Some VSM Symbols

Symbols from
http://www.epa.gov/lean/
environment/toolkits/envir
onment/app-a.htm

C/T	Cycle time
C/O	Changeover time
⚠	Inventory
🚚	Truck shipment
🏭	External sources (suppliers, customers, etc.)
↘	Electronic information flow
➡	Movement of production material
⊐	Supermarket (a controlled inventory of parts)
C	Withdrawal (pull of materials, usually from a supermarket)
▱	Production kanban (card or device that signals to a process how many of what to produce)
⌐▽⌐	Signal kanban (shows when a batch of parts is needed)
✴	Kaizen starburst (identifies improvement needs)

41

Figure 10.3 Value stream mapping symbols

10.5 Kaizen Event Planning

The kaizen event is based on the 5S methodology (Figure 10.4). We first Sort and remove unneeded or broken tools, parts and supplies from the targeted area. Then we Set in Order, putting all the remaining items in their place, creating shadow boards, etc. We then Shine or clean the area. Next, we Standardize, making sure that everyone does things the same way. Finally, we Sustain, in which we maintain the gains and make 5S a habit so that the area stays organized.

Chapter 10 Implementing Lean © 2021 University Training Partners

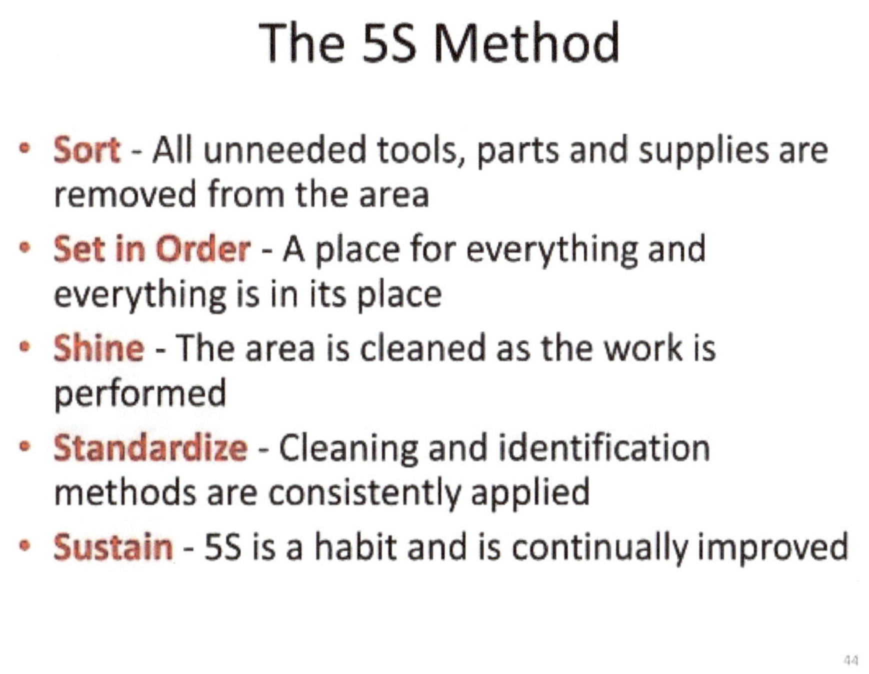

The 5S Method

- **Sort** - All unneeded tools, parts and supplies are removed from the area
- **Set in Order** - A place for everything and everything is in its place
- **Shine** - The area is cleaned as the work is performed
- **Standardize** - Cleaning and identification methods are consistently applied
- **Sustain** - 5S is a habit and is continually improved

44

Figure 10.4 The 5S method

Successful kaizen events take careful planning. (Martin, 2007) The list of planning activities (Figure 10.5) sounds a lot like the activities performed in the Define phase in a Six Sigma project. We can leverage the existing project charter, SIPOC diagram, team charter and stakeholder analysis templates for this event.

Event Planning Activities

- Choose project
- Select leadership team
- Decide on scope
- Decide on duration and dates
- Identify resources needed
- Complete a Kaizen Event charter
- Choose participants
- Develop communication plan
- Develop and perform preliminary training
- Prepare for team development
- Complete team charter

Source: Martin & Osterling, **The Kaizen Event Planner**, CRC Press, 2007

50

Figure 10.5 Kaizen event planning activities

The kaizen event itself is characterized by having a cross-functional team work on identifying the current state of the process, identifying waste, and then mapping the future state. The kaizen event is also referred to as a "blitz" to emphasize the short duration and aggressive objectives of the exercise. The event is considered complete only when there is full implementation of changes, the workforce has been retrained, and mechanisms for sustainability have been implemented.

It is very important to document the current state with photos, a spaghetti diagram, or process flow chart and performance metrics. The "before" is a vital part of the improvement story. After that, identify the value-added versus non-value-added pieces of the process, and identify the root causes of the waste. The team will do this by using 5 Whys, fishbone diagrams or Pareto charts.

After identifying the root causes of the problems, the team can use a PACE diagram to prioritize the improvement options the team develops, as shown in Figure 10.6. We place each option on the grid according to how beneficial it will be versus how difficult or easy it will be to implement. Then, we draw contours that map to:

- P = priority – the team will do tackle these improvement items first
- A = action – the team will implement these items next
- C = consider – the team will consider these items later and
- E = eliminate as the payoff decreases and the difficulty increases.

Note that the PACE diagram is a more refined version of the payoff matrix introduced in Section 9.9.

Executing the Event

- ## Brainstorm and prioritize improvement options
  - ### Use a Priority-Action-Consider-Eliminate Diagram

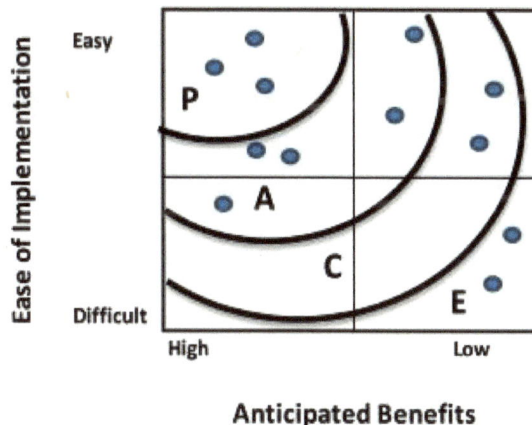

Source: Martin & Osterling, The Kaizen Event Planner, CRC Press, 2007

Figure 10.6 PACE diagram

Once the improvements have been decided upon, the actual kaizen event can be held. During the event, the improvements are designed, tested, and implemented. It is at this point where the team gets its hands dirty and performs the sorting, setting in order and shining. After the intense burst of activity, there will be a kaizen event report, with before and after pictures, and sustainability plan, much like a control plan in a Six Sigma project. In addition, the report will include a punch list of items that still need to be cleared up over the next 30 days, and a listing of parking lot issues that were captured in the meetings. A final team presentation is given, and then, as in a Six Sigma project, the team is recognized and disbanded.

10.6 Mistake Proofing

Mistake proofing is also known by its Japanese name *pokayoke*. The technique is always handy, but especially useful in preventing defects when they are rare, and effective in eliminating random, non-systematic errors. In other words, mistake proofing is good for preventing rare but serious events, and for preventing random human error.

For example, I once worked in a factory that manufactured electrical switches for the automotive industry. One of the switches was a simple plunger switch that turned the interior lights on in the car when the door was opened. The switch was composed of two metal shafts, one of which fit inside the other. In one operation, the larger diameter shaft was put into a machine that crimped a flange on the shaft. An operator stood in front of the machine, placing one switch in at a time, pressing the buttons on either side. The machine came down on the switch and created the flange, repeatedly, for eight hours a day. The operator knew how to do her job, but occasionally, for no reason, she would put the switch in the machine upside down, and the flange would be put on the wrong shaft. This mistake made the switch inoperable. She might do this a few times an hour, or only once a shift, but it did happen. This was a random, human error and not a process error that could be caught by a statistical process control (SPC) chart, for example.

In this situation, more training would not be effective since the operator knew what to do. The mistake was just an error of inattention that happened randomly and rarely. Because of the importance of the switch, the resulting product had to be 100 % inspected because the production manager was never sure when or if she would make the error. The extra inspection took a lot of time and effort. To improve the situation, we decided to eliminate the error by design. We changed the fixture on the machine to incorporate a slot that the switch subassembly slid into. The slot accommodated the smaller shaft, but the larger shaft could not fit. We thus eliminated the possibility of a switch being placed in the machine upside down and completely eliminated the error. The 100% inspection was no longer necessary. It cost about $5 to make the new fixture in the tool shop of the plant.

There is a classic definition of mistake proofing, which is the complete elimination of errors, but over the years, mistake proofing has been expanded to include controls or warnings. With this version, having a machine shut down after a mistake is made, or a light or alarm sound when an error occurs is considered mistake proofing. Check lists would also fall into the control category. All these devices are effective if the warning is heeded, or if the check list is adhered to. However, the gold standard of course is to have a design in which the mistake will not occur in the first place.

Classic examples of mistake proofing include diesel and unleaded fuel nozzles at the gas station. You cannot put diesel fuel into your gas-powered car, the nozzle just won't fit. The same was true with leaded and unleaded gasoline. This is an elimination of a defect by design.

In cars, the key release and gear shift are a mistake-proofed system. The gear shift must be in park before the key can be removed, or the brake must be depressed before the car can be taken out of park. These are error elimination techniques by design.

In contrast, there is an oil warning light on the dashboard that will signal when the oil level is low. Upon seeing this signal, a driver should immediately pull over to put oil in the engine. Alas, this warning light can be ignored, as someone with teenage drivers can tell you.

In the hospital, machines tracking oxygen levels, heart rhythms, and IV levels produce audible signals when action is needed. However, it has been shown that "alert fatigue" can set in when there are too many signals occurring too frequently. Safety seals on packages are another mistake proofing system. If the seal is broken, it is an alert to the consumer that the project may be tainted. This is not an elimination of a defect, but a warning.

10.7   Quick Changeover

Changeover reduction can be used to reduce or eliminate waiting waste and increase machine availability. Changeover reduction is also known by the acronym SMED, which stands for *single minute exchange of dies.*

In the hotel industry, housekeeping must turn over a room to prepare for the next guest. In manufacturing, maintenance must change over a production line when it is switched from making one product to the next. If changeover times are reduced, production lines can be run with smaller batches, which reduces inventory waste and allows the facility to match its output to customer demand. If more time is spent in production and less in changing over, the facility can also increase its capacity (Figure 10.7).

Changeover time is defined as the total time from the *last unit of production* to the *first unit of good production at full speed.* The goal is to reduce changeover time so that planning is nimbler, and throughput is increased.

Changeover Reduction

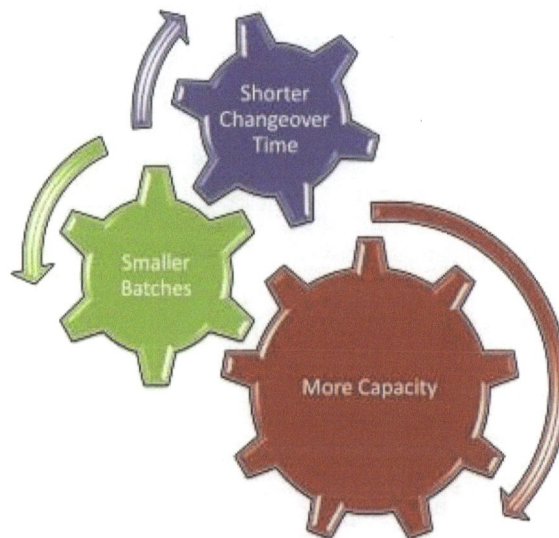

Figure 10.7 Changeover reduction

Dr. Shigeo Shingo was developed single minute exchange of dies (SMED) while working with an auto manufacturer. The company dealt with changeovers all the time. Let's say for example, that the manufacturer was faced with switching out a door line in the pressed metal plant to stamp doors for a different model. How could this changeover time be reduced? By using Shingo's methods, the company was able to reduce the changeover time from hours to less than ten minutes – thus the "single minute" name. To accomplish this dramatic reduction, the team converted *internal work* to *external work*.

Internal work is counted as all the tasks performed while the line is shut down. The external tasks are performed either before the line is shut down or after the line is up and running again. By performing a detailed work break down, the team was able to convert internal tasks to external tasks, doing things ahead of time, or after the fact, so that the actual downtime and internal work was minimal.

As an example, consider a pit crew in Formula 1 or NASCAR racing. The crew's job is to make the actual pit stop as sort as possible — when the car is in the pit, it is not racing. To minimize downtime, the pit crew does as much as possible ahead of time to get ready for the internal tasks. They have specialty tools designed to perform their tasks. They are not using a lug wrench and then trying to find all the lug nuts once they are putting a new tire on. Specialty equipment and quick fasteners allow the tires to be changed in a matter of seconds. Each person on the crew has a specific job, and the crew practices daily to be able to perform their tasks effectively and efficiently.

The pit stop mentality is needed when reducing changeover time on the manufacturing floor, or in the hospital operating room, or in a hotel. We list each task, write out the detailed act breakdown, and then figure out how to move tasks from internal to the external.

Unlike other tools used in Lean, SMED may require some significant capital investment, such as specialty tools and fasteners designed to minimize down time. The goal is to re-design, re-engineer, and re-tool.

Chapter 11 Lean Calculations

Although Lean is not as math-intensive as Six Sigma, there are production-oriented calculations that can help a team establish baselines and measure process improvements. This chapter will cover takt time, overall equipment effectiveness, rolled throughput yields and Little's Law.

11.1   Takt Time

Lean calculations can help a facility to set the pace of production and to measure process efficiency. *Takt* is a German word that means "beat" or "cadence." The takt time indicates the pace that products or services should be completed to meet customer demand. Takt time will vary according to customer demand.

To calculate takt time, the available work time in minutes is divided by the customer demand.

$$\text{Takt time} = \frac{\text{Available work time}}{\text{Customer demand}}$$

For example, within an 8-hour shift, there are required breaks and a lunch period, so the actual available work time will be less than the full 8 hours. For this example, let's assume that two 15-minute breaks and a 30-minute lunch period occur each shift, resulting in 420 available minutes for active work. If there are 50 customer orders to be filled today, then, working at a steady pace, an order needs to be to completed every 8.4 minutes to meet customer demand.

$$\text{Takt time} = \frac{420 \text{ minutes}}{50 \text{ orders}} = 8.4 \text{ minutes/order}$$

This calculated takt time can be used to properly staff a production line or a service desk. For example, if one service desk worker can process a customer order in 17 minutes, we can use the takt time to determine how many workers need to be assigned to the desk to meet customer demand for that day. Recall that the takt time was 420/50 = 8.4 minutes/order. The cycle time is equal to 17 minutes, but we need to finish one every 8.4 minutes, we will need to have (17 ÷ 8.4) or about 2 workers staffed at the desk to meet customer demand. If the takt time is very small compared to our cycle time, we will have to add capacity, or redesign the process to increase throughput.

11.2　Overall Equipment Effectiveness

Another useful Lean metric is overall equipment effectiveness, or OEE. It is a measure of overall utilization of facilities, time and materials in a manufacturing operation based on scheduled run time. It is calculated from the terms Availability, Performance and Quality. (Figure 11.1)

Calculating OEE Components

Availability = (Available Time - Downtime)/ Available Time

Performance = Actual Production Rate/ Designed Production Rate

Quality = Good Pieces/ Total Pieces

Figure 11.1 OEE components

Availability is calculated as the ratio of available machine time less down time, divided by available time.

Performance is the ratio between the actual production rate and the designed production rate.

The *Quality* metric is yield, or the number of good pieces divided by the total pieces produced.

We want to maximize overall equipment effectiveness, obviously. We can do that by addressing what are called the Six Big Losses (Figure 11.2). The first two losses are equipment breakdowns and set up and changeover time. Together, these two losses affect Availability.

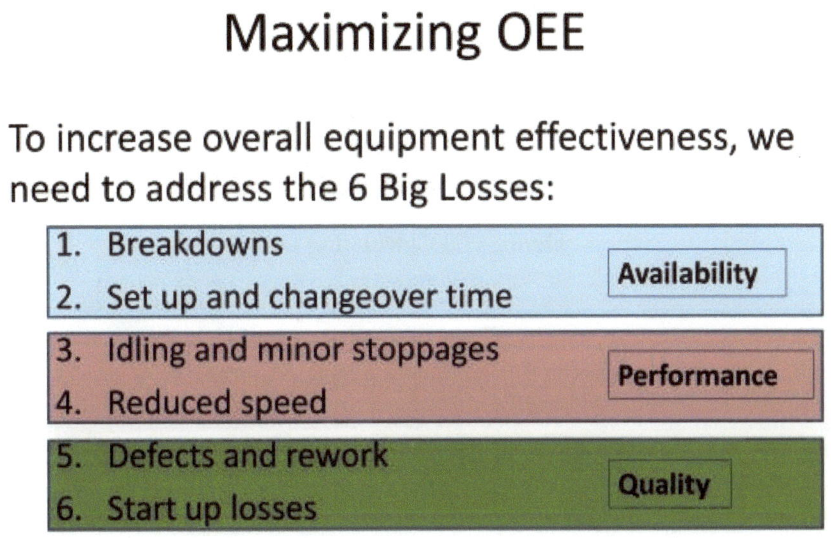

Maximizing OEE

To increase overall equipment effectiveness, we need to address the 6 Big Losses:

1.	Breakdowns	**Availability**
2.	Set up and changeover time	
3.	Idling and minor stoppages	**Performance**
4.	Reduced speed	
5.	Defects and rework	**Quality**
6.	Start up losses	

Figure 11.2 The Six Big Losses

Next, we have the losses due to minor stoppage and idling machines, as well as reduced machine speed. These losses adversely affect the matrix we call Performance.

Finally, the 5th and 6th Big Losses are defects and rework, and scrap due to machine start up. These losses fall under the Quality umbrella. If we reduce the Six Losses, we will increase OEE.

To calculate overall equipment effectiveness, we multiply Availability, Performance and Quality, as shown in Figure 11.3.

Calculating OEE
OEE = **Availability** x **Performance** x **Quality**
OEE = **A** x **P** x **Q**

Figure 11.3 Calculating OEE

Consider a truck rental company that has one computer dedicated to processing credit card transactions. The facility is open 16 hours per day, 5 days a week, and the computer happened to be offline 3 hours last week. The computer is capable of processing 100 agreements per hour. There were 7000 transactions processed last week with a total of 250 transaction errors.

Putting this data to work, we can calculate Availability, which is available time minus downtime, divided by available time. The available time for a week is 80 hours, calculated as 16 hours per day times 5 days. The computer was down for 3 of those hours, so the Availability was 0.9625, or 96.25%, as shown below.

$$\text{Availability} = (\text{Available Time} - \text{Downtime})/ \text{Available Time}$$
$$\text{Availability} = ((16 \times 5) - 3)/(16 \times 5) = 77/80 = 0.9625$$

The Performance is the ratio of the actual production rate versus the designed rate. Seven thousand transactions were processed last week during the 77 hours that the computer was online. The system is designed to process 100 transactions per hour. Performance is running at 90.91%.

$$\text{Performance} = \text{Actual Production Rate}/ \text{Design Production Rate}$$
$$\text{Performance} = (7000\,\text{units}/77\,\text{hours})/(100\,\text{units/hour}) = 0.9091$$

Quality is calculated as good pieces produced divided by the total produced. There were 250 errors, in the 7000 total transactions, producing a yield of 96.4%.

$$\text{Quality} = \text{Good pieces/Total pieces}$$
$$\text{Quality} = (7000 - 250)/7000 = 0.964$$

Chapter 11 Lean Calculations © 2021 University Training Partners

Because each of the three metrics that are multiplied together in this case is less than one, the OEE will be less than the lowest metric of 87.5%. In fact, OEE is equal to 0.844, or 84.4%.

$$OEE = A \times P \times Q$$
$$OEE = 0.9625 \times 0.9091 \times 0.964 = 0.844$$

In this example, OEE is limited by the lowest metric, Performance. Even if we improved availability and quality to 100%, OEE would still only be able to increase to 0.9091. The system is designed to handle 100 transactions per hour, but we are only processing 91. This may be a function of our customer demand, in which case our system is overdesigned, or it could be that only 7000 transactions could be processed because of minor stoppages, or workers are processing transaction slower than designed due to lack of training, for example.

Take a few minutes to calculate the OEE for a metal fabrication shop given the following information:

A metal fabrication shop produces welded subassemblies. The facility is open for 16 hours a day, six days a week. During each eight-hour shift, there are two 15-minute breaks and a 30-minute lunch break. The welding process is designed to produce 45 subassemblies per hour. A total of 500 units were produced yesterday, of which 55 were defective. The line experienced a breakdown yesterday, resulting in 2 hours of downtime. Calculate the OEE for the welding process.

How did you do? The answer appears below.

The available work time each day is equal to 16 hours less the breaks and lunch periods. There are four fifteen-minute breaks and two lunch periods across the two daily shifts. Note that we must first convert the breaks to fractional hours.

$$\text{Availability} = (\text{Available Time} - \text{Downtime})/ \text{Available Time}$$
$$\text{Availability} = (16 - (4 \times 0.25) - (2 \times 0.50) - 2.0)/(16 - (4 \times 0.25) - (2 \times 0.50))$$
$$= 12/14 = 0.857$$

The actual production rate is equal to the number of pieces produced divided by the actual work hours.

$$\text{Performance} = \text{Actual Production Rate}/ \text{Design Production Rate}$$
$$\text{Performance} = (500 \text{ units}/12 \text{ hours})/(45 \text{ units}/\text{hour})$$
$$= (41.67 \text{ units}/\text{hour})/(45 \text{ units}/\text{hour}) = 0.9259$$

The quality rate is the number of good pieces produced divide by the total number of pieces produced.

$$\text{Quality} = \text{Good pieces}/\text{Total pieces}$$
$$\text{Quality} = (500 - 55)/500 = 0.89$$

$$\text{OEE} = A \times P \times Q$$
$$\text{OEE} = 0.857 \times 0.9259 \times 0.89 = 0.706$$

Here, the availability is the limiting factor in the OEE calculation due to the two-hour downtime. The root cause for the downtime needs to be found so that it can be prevented in the future. The next step in raising OEE is to understand and eliminate the root causes for the defective pieces that were produced.

11.3 Throughput Yield (TY)

Throughput yield (TY) is the percentage of good pieces that are produced by a process. We can estimate the throughput yield (TY) of a process by using a Poisson approximation to the binomial distribution. If we know the average number of defects per unit (dpu) produced by a process, we can estimate the TY by using the Poisson formula that corresponds to $\Pr(X = 0)$:

$$TY = e^{-dpu}$$

Here, "e" is the irrational constant 2.71828... You will find a "e" key on a scientific calculator, and it can be calculated using the =exp() formula in Excel. For example, to find the value $e^{-1.4}$ using Excel, we would type =exp(-1.4). This Poisson approximation works well when the total number of opportunities for defects in the unit is greater than 100, and the probability of a defect is low, $p < 0.10$.

Here is an example of the throughput yield (TY) calculation using the exact probability as well as the approximation. A hotel performs a housekeeping audit of 15 guest rooms. The audit uses a checklist of $n = 10$ items which are marked as complying or non-complying. (Note that the total number of opportunities for defects for each room is equal to ten, which is less than the approximation criterion of 100.) We will proceed with the approximate and exact TY calculations to note the difference between the two methods.

The audit of 15 rooms finds a total of 18 errors. The number of defects per unit (dpu) would then be $18/15 = 1.2$. On average, each room will have 1.2 non-compliance issues.

To use the Poisson approximation to the binomial, we set the Poisson mean equal to the dpu. The TY, or the probability of finding zero defects in a room audit, is estimated as:

$$TY \cong e^{-1.2} = 0.30$$

To find the exact throughput yield, we must calculate the probability of a checklist item being out of compliance in a single room. Assuming the $n = 10$ checklist items are independent, the defects per opportunity can be calculated as:

$$dpo = dpu/n = 1.2/10 = 0.12$$

We use this dpo as the probability of a checklist item being out of compliance for a single room. Note here that $p = 0.12$, which does not meet the approximation criterion of $p < 0.10$.

The probability of an opportunity being in compliance is $(1 - 0.12) = 0.88$, based on the complement rule. Now we have all the information needed to calculate the exact TY using the binomial distribution.

The throughput yield (TY) will be equal to the probability of a single room having no defects. Using the binomial distribution with n = 10 and p = 0.12, the probability of finding no defects is $0.88^{10} = 0.279$.

The TY = 0.30 found through the Poisson approximation is 8.0% higher than the exact TY. Note that the number of defect opportunities per unit was 10, which did not meet the approximation criterion, and that the probability of a defect was greater than 0.10.

The Poisson approximation will improve as the number of checklist items increases. For example, let the number of checklist items be equal to 100, with 18 defects found in the 15 rooms. As before, dpu is equal to 18/15 = 1.2 and the Poisson approximation yields

$$TY = e^{-1.2} = 0.30.$$

The defects per opportunity which is used as the probability in the binomial distribution is now:

$$dpo = dpu/n = 1.2/100 = 0.012$$

For a single room, the probability of finding a defective item on the checklist is 0.012. Using the complement rule, we can state that the probability of an opportunity being in compliance is (1 - 0.012) = 0.988.

The exact binomial probability of a room having no defects in a room audit is then equal to $0.988^{100} = 0.299$, which is also the throughput yield.

Note that the approximation is much improved since n ≥ 100 and p < 0.10.

11.4 Rolled Throughput Yield (RTY)

We can assess the overall throughput of a multi-step process by considering the first-time yield of each process step. The overall throughput is called the rolled throughput yield (RTY), and it is an excellent barometer of the capability of a process.

Assume that parts are inspected after each step in the process. The first-time yield (FTY) of a process step is the net percent output of the process. It is calculated as the net output (the number of piece input, less scrap and rework) divided by the total input. It is expressed as a decimal, or a percent:

$$FTY = (input - scrap - rework)/ (input)$$

Because reworked pieces can continue through the process, the number of pieces that are fed into the next process down the line is calculated as:

$$Input\ for\ next\ process = input - scrap$$

The rolled throughput yield (RTY) is then calculated as the product of the first-time yield for each of the process steps.

For example, consider a process with four consecutive steps, A, B, C, D. The production information for each step is shown below.

In Step A, 100 pieces are processed, resulting in 5 scrapped pieces and 10 pieces that are rejected, but can be reworked.

Step A: 100 input, 5 scrap, 10 rework

FTY A= (input A – scrap A – rework A)/ (input A)

$$FTY\ A = (100 - 5 - 10)/100 = 0.85 = 85.0\%$$

We assume that the rework is processed and put back into production. Therefore, the number of pieces input to step B is equal to the pieces started by Step A, less the scrap pieces produced in A. The input to step B can be calculated as:

Input to Step B = input A – scrap A = 100 – 5 = 95

In Step B, the 95 pieces processed result in 4 scrapped pieces and 2 requiring rework.

Step B: 95 input, 4 scrap, 2 rework

FTY = (input B – scrap B – rework B)/ (input B)

$$FTY = (95 - 4 - 2)/95 = 0.937 = 93.7\%$$

Input to Step C = input B – scrap B= 95 – 4 = 91

Step C: 91 input, 0 scrap, 5 rework

FTY = (Input C – scrap C – rework C)/ (Input C)

FTY = (91 – 0 - 5)/91 = 0.945 = 94.5%

Input to Step D = input C – scrap C= 91 – 0 = 91

Step D: 91 input, 1 scrap, 3 rework.

FTY = (input D – scrap D – rework D)/ (Input D)

FTY = (91 – 1 – 3)/91 = 0.956 = 95.6%

The rolled throughput yield (RTY) is then calculated by multiplying the FTY for each of the 4 process steps:

$$RTY = \prod_{i=1}^{4} FTY_i$$

$$RTY = (0.850)(0.937)(0.945)(0.956) = 0.720 = 72.0\%$$

The rolled throughput yield for the entire process is 72%.

If we want to compare rolled-throughput yield of several competing process designs, for example, we can "normalize" the metric to remove the influence of varying levels of complexity among designs. To do this we calculate the geometric mean of the first-time yield of each process to obtain the normalized rolled-throughput yield, NRTY. For a process with m steps, we multiply each FTY and then take the mth root:

$$NRTY = \sqrt[m]{\prod FTY_i}$$

For example, we can calculate NRTY using the data from the previous example by taking the fourth root of the RTY.

$$NRTY = \sqrt[4]{\prod FTY_i} = \sqrt[4]{0.850 \times 0.937 \times 0.945 \times 0.956} = 0.72^{1/4} = 0.921$$

By normalizing rolled throughput yield, we can then compare the performance of the process to others with varying numbers of steps.

When performing the calculations in Excel, we need to put parentheses around the fraction in the exponent. For example, the previous calculation would be entered as =(0.850*0.937*0.945*.956)^(1/4).

11.5   Little's Law

Little's Law is a mathematical relationship that is used in queueing theory and manufacturing flow calculations. It very succinctly relates the number of items waiting for processing, the wait time of each item, and the number of items arriving per time period. Letting

L = average number of items waiting in the queue in the system

W = average waiting time in the system for an item

λ = average number of items arriving per unit time,

we can write the law as

$$L \; = \; \lambda W.$$

This equation is pretty remarkable in its applicability. The relationship requires that the system operate at a steady state, or in other words, the system must run in a predictable fashion. But that is really the only requirement.

Little's Law will hold no matter if we have two tellers or six tellers working at the bank, whether we have one line or multiple lines of customers, and no matter what type of statistical distribution the service time follows. It also does not matter under what distribution the customers arrive. Little's Law will still hold whether we are doing first-in-first-out or last-in-last-out servicing. Although the law is a very powerful and useful formula, it is simple enough to be calculated on the back of an envelope (although there is a Little's Law calculator in the course companion spreadsheet).

The applications of this law are wide ranging. We can estimate the number of diners waiting for a table on a Friday night at a popular restaurant, the number of patients waiting at a Covid-19 testing center, or the number of people waiting in line for a rollercoaster ride at an amusement park. In each case, the estimated number of people waiting in line is a function of the duration of the service (cycle time) and the number of people arriving in the queue per minute. Using these figures, planners can make staffing and waiting area decisions to accommodate demand.

We can also define the law's variables in terms of manufacturing, with:

L = work in process inventory (WIP)

W = process lead time (how long a new unit will have to wait to get worked on)

λ = completion rate in terms of items/time period. (Hopp, 2001)

By doing some quick algebra, we can also calculate process lead time (W) by taking the number of pieces waiting to be processed (L), or the number of people ahead of you in the bank line,

and dividing by the completion rate (λ). The completion rate is the time it takes for a piece to be processed by the machine, or the amount of time it takes a teller to complete a customer's transaction.

$$W = \frac{L}{\lambda}$$

In this form, Little's Law can be used to calculate how long a new diner will wait to get seated, how long a patient will wait for a Covid-19 test, or how long the wait for the rollercoaster ride will be.

Chapter 11 Lean Calculations © 2021 University Training Partners

Chapter 12 Training Employees

Very often the Black Belt's responsibilities extend past working on improvement projects. Many Black Belts have a primary job responsibility of training developer, or trainer. Besides directing formal training programs, Black Belts also perform on-the-spot training for their teams, explaining new tools as they are needed, or giving overviews of the DMAIC steps. Thus, Black Belts need to know how to train large and small groups effectively.

12.1 Four Principles of Adult Learning

There are four basic principles of adult learning. The first is *readiness*. Learners must perceive a benefit to the training to be receptive to it. The second principle is *experience*. Trainers must incorporate the learners' prior experience into the training: training should build on what learners already know. The third principle is *autonomy*. Learners must be actively involved in their learning. The fourth principle is *action*. Learners must see how they can apply their new skills immediately.

These principles bear out in student reviews of training sessions. Typical student responses for a positive training experience include:

- There was a lot of participation
- Training was fun
- I saw how I could use these tools at my job
- The materials were clear
- The atmosphere was casual
- I enjoyed interacting with the other participants

Examples of common negative feedback are listed below:

- The class moved too fast
- This training has nothing to do with my job
- The instructor was intimidating
- There was a lot of wasted time.

- There was little interaction
- I didn't get to test my knowledge

12.2 Training Tips

We all have vast classroom experience as students. However, training is not the same as "doing school." Education and training and education are not the same: the purpose of training is to instruct students on performing specific skills.

Adult trainees do best when they own their learning and are not made into passive recipients of knowledge. (I suppose this can apply to children as well!) Adults learn best through experience and need to be shown how the knowledge being presented can be directly applied to their jobs. The "Forgetting Curve" was developed by German psychologist Hermann Ebbinghaus in the late 19th century, shown in Figure 12.1. (Glaveski, 2019) The curve demonstrates the negative effect of not applying what we have learned. After just six days, students forget 75% of what they have learned if the new knowledge is not applied.

The Forgetting Curve

If new information isn't applied, we'll
forget about 75% of it after just six days.

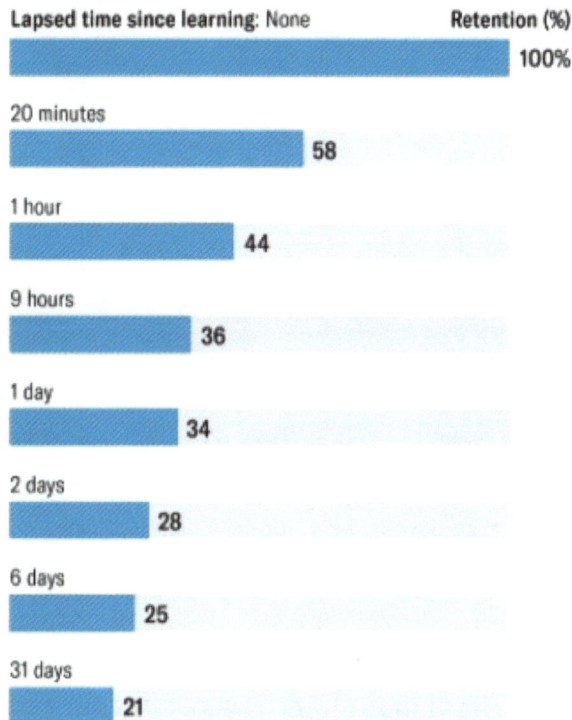

Lapsed time since learning: None	Retention (%)

Lapsed time since learning: None — 100%

20 minutes — 58

1 hour — 44

9 hours — 36

1 day — 34

2 days — 28

6 days — 25

31 days — 21

Source: Hermann Ebbinghaus   ♡ HBR

Figure 12.1 Ebbinghaus's Forgetting Curve

Learning should also be enjoyable! The Black Belt has the opportunity to break away from the example set by those interminable college lectures and think about how a concept can be presented in novel ways. Techniques might include simulations, games, role play, group work, or class discussions.

It is said that you don't truly know a subject until you can teach it to someone else, and Black Belts will discover quickly that this is true. It takes an incredible amount of preparation to develop and present a successful training program. Being able to answer student questions and give examples while standing in front of a classroom full of students is the measure of an instructor's understanding of the material. Instructors can leverage this phenomenon by asking the students to perform teach-backs, in which they present the material they have learned to a group of students. Preparing for a teach-back will expand and solidify students' understanding. (Stolovitch, 2002)

As quality practitioners, Black Belts see activities through the lens of a process. Training is no different here: we are right in treating training as a process and not a one-time event. Just because the classroom time is over does not necessarily mean that the learning is complete. Instructors should follow up with students to check how well they have absorbed their new skills and redirect them if necessary.

12.3   Instructional Design Models

Careful design of the training program can help create a positive experience for your students. There are several instructional design paradigms in the adult education literature. The well-known ADDIE model for training development, for example, uses the steps Analysis, Design, Development, Implementation and Evaluation. (Figure 12.2)

Figure 12.2 ADDIE instructional design model

The five-phase ADDIE model gives instructional designers a road map for creating successful training programs. In the Analysis phase, designers identify the learners and their characteristics, state the new behavior desired, identify learning constraints, list delivery options, and plan using project timelines, among other tasks.

In the design phase, a blueprint of the training is created. The learning objectives and assessment instruments are identified. Then, the designers plan the course content, and group and individual exercises.

In the development phase, the designers create storyboards and bring the elements of the design blueprint to life.

In the implementation phase, trainers deliver the course to the learners.

The evaluation phase is then conducted to make sure the training design met the requirements, and that the learning objectives were met. At this point, the design team can return to the analysis phase to make changes in the program.

The ADDIE model shares many similarities with systematic approaches used in quality. For example, we can use Dr. Deming's PDCA cycle, Plan-Do-Check-Act as our framework for developing a training event, displayed in Figure 12.3. Note that the PDCA cycle does not end, meaning instructors are constantly improving their training materials and delivery.

PDCA for Trainers

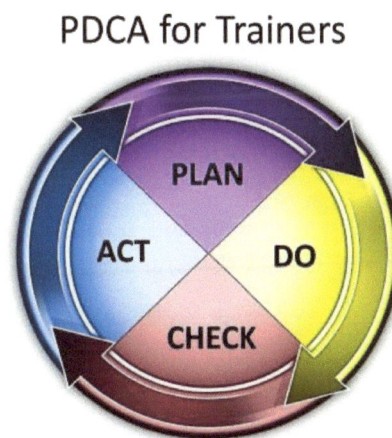

Figure 12.3 PDCA model

In the Plan phase of training development, the trainer needs to answer basic questions.

- What will you be teaching?
- Will it be a single tool, or an entire Green Belt program?
- Why is the training needed?

There must be a specific purpose for the training. What are the desired outcomes, as in, what should the students be able to do after the training? What are the learning objectives?

Next, the trainer goes through a personal assessment. As an instructor, what are your strengths and weaknesses in terms of presentation skills, knowledge gaps, computer skills, and comfort with various group activities? This should help you design a program that will be effective. It might be that you realize that you need to improve your skills in using a smart board, for example. Until you are comfortable with that technology, you might plan to deliver your lessons another way. It might be that the thought of leading a group simulation with 30 students intimidates you. In the meantime, build your confidence by leading smaller groups.

More questions need to be answered in the Plan phase. Who will you be training? Are they engineers, line workers, pharmacists, or accountants? The audience will dictate the level and depth of the presentation. Why does the group need the training? What is in it for them? If you can see no immediate benefit to the training, they will not be able to either. What are the students' specific interests, strengths, and weaknesses in terms of interpersonal skills, knowledge of the company, math skills or computer skills?

How many students will be in class? If there will be ten to twenty learners, that is an appropriate size for group activities. If there will be fifty students, the instructor may have to change the approach to some of the exercises. Will the training be performed online or in person? Clearly, different instructional formats are more suited to various desired outcomes.

The next phase in PDCA is Do. What content will you deliver? Will it come from a textbook, or from online resources? How will you present the material? Will it be through PowerPoint lecture, videos, guest speakers, simulations, etc.?

At the Check phase, you will need to incorporate assessments in the training design to make sure the learning objectives have been met. How will you go about assessing that learning has taken place? You may choose to check understanding through individual exercises or group work, homework assignments, exams, or student feedback. You may also decide to track how students perform on the job after the training.

The last step in the PDCA cycle is Act. Here, you deliver the training, and then reassess the course considering went well and what could be improved. This assessment can come from personal observation, student feedback forms, and test results. And once the training is completed you can reassess the training mode, exercises, length of the training: in short, everything.

We then go back to the Plan phase, moving through the PDCA cycle to continuously improve our training materials and delivery.

For students, their learning also follows a PDCA cycle. To optimize learning, students should (Glaveski, 2019):

1. Learn the core of the knowledge they need (Plan)
2. Apply the learning to a real-world situation immediately (Do)
3. Receive immediate feedback (Check)
4. Refine their understanding (Act)
5. Repeat

Glaveski (2019) presents a training approach that he has named Lean Learning. This paradigm draws from the principles of Lean manufacturing. The Lean approach starts with the use of the Pareto Principle. Eighty percent of the new behavior of skill can be accomplished by learning 20% of the available tools. For example, 80% of Six Sigma projects can be successfully led by using 20% of the available Six Sigma tools. Why put Green Belts through weeks of intensive training when they could accomplish the same improvements with one week of training on the most frequently used tools?

Another Lean training principle is to deliver the instruction at the time and place it is needed. This calls for on-the-spot training of tools as the team needs them or developing a suite of mini courses that train on a single tool. Finally, skills can be sustained by providing ways for students to receive support after the training session via video conferences or email correspondence.

Black Belts may be tapped to design a training module that teaches students how to use a specific tool. Even with this small-scale assignment, the ADDIE, PDCA or Lean approach should be used to develop the training session. Black Belts may even be tasked with designing and delivering entire Yellow and Green Belt programs, which can be a daunting task. For those seeking a quick start guide for course design, the text *Telling Ain't Training* by Stolovich and Keeps is a great resource. Sharon Bowman has written *The Ten-Minute Trainer* which is an interesting repository of training activities. Both texts are listed in the Bibliography.

Chapter 13 Designing for Six Sigma

Instead of improving an existing process or product, sometimes a team is given the task of designing a product or process from scratch. Six Sigma principles can be applied to the design process by using a modified DMAIC cycle called Design for Six Sigma (DFSS), shown in Figure 13.1. Design for Six Sigma, or DFSS, uses the DMADV cycle, which is made up of Define, Measure, Analyze, Design and Verify phases. We will learn some tools that can be helpful when designing a new product or process.

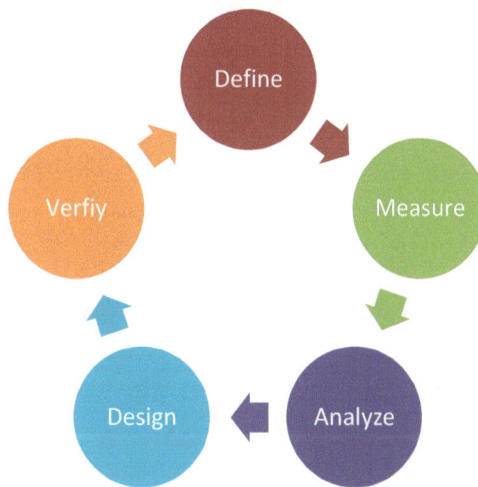

Figure 13.1 The Design for Six Sigma DMADV Cycle

13.1 Affinity Diagrams

Affinity diagrams are used to help teams think about the types of features a product could have, or what issues need to be considered when a new process is being implemented.

Affinity diagrams are a great first step when the team is faced with a large, ill-defined problem, or a blank sheet of paper on the drawing board. The team will brainstorm ideas on features or issues that need to be addressed and then group these ideas into categories. These categorized ideas begin to give structure to the problem.

For example, in one of my past jobs I was a professor at a small university. Our university offered engineering technology degrees, which are more practical and hands-on than engineering degrees. The school had been petitioning the state to allow it to offer engineering degrees, but the requests were repeatedly denied, no doubt due to the influence of the big, prestigious engineering school in the state. Finally, the university was granted permission to offer engineering degrees. This was quite a victory, but then the administration and faculty all looked at each other and said, "Well, now what?"

The accrediting bodies for engineering technology and engineering are different, as are the requirements for professors in terms of education and research. There are also different standards for the course offerings, lab requirements, and on and on.

The administration convened the engineering technology faculty – there were about 100 of us. We were broken up into groups and asked to brainstorm all the things we would need to consider as we transitioned from an engineering technology school to one that also offered engineering degrees. We brainstormed and wrote hundreds of ideas on flip chart paper that went around the large auditorium. We identified issues like tenure requirements, salary differentials, curriculum design, calculus requirements, admission criteria, classroom space, and on and on.

After that initial meeting, all the ideas were collated, and similar ideas combined. The faculty was then sent a listing of ideas that were combined into what was thought as natural groupings: all the ideas that dealt with facilities, or curriculum, or hiring, or admissions criteria, etc.

The administration then assigned subcommittees to take over each natural grouping to begin to work through the transition issues. The university had a large marshmallow of a problem – how to create a new engineering school – and it was able to give the problem by using a modified version of the affinity diagram.

To create an affinity diagram, the team will gather ideas written on post-it notes from a brainstorming session, or information collected from customer feedback. The team will then begin to take these ideas and move them into natural categories, *without talking*. The silence is to prevent one person taking over the sorting process. After the team is finished sorting the

ideas into categories, it then decides together what each category should be called. The resulting groupings form an affinity diagram, in which ideas and issues for a large project are organized into categories. The team now can begin to tackle the problem category by category.

13.2 Quality Function Deployment

Another Design for Six Sigma tool is the House of Quality. The House of Quality is created when the team uses Quality Function Deployment or QFD, a structured process for planning a new process or product. Both QFD and the House of Quality help us translate customer needs into features and specifications for the new product. Using QFD shortens the design time, reduces design costs, and results in a product that pleases the customer.

The House of Quality is a set of matrices that are put together to look like a house, as shown in Figure 13.2. The areas of this house include the left and right wings, the roof and attic, the center, and the basement.

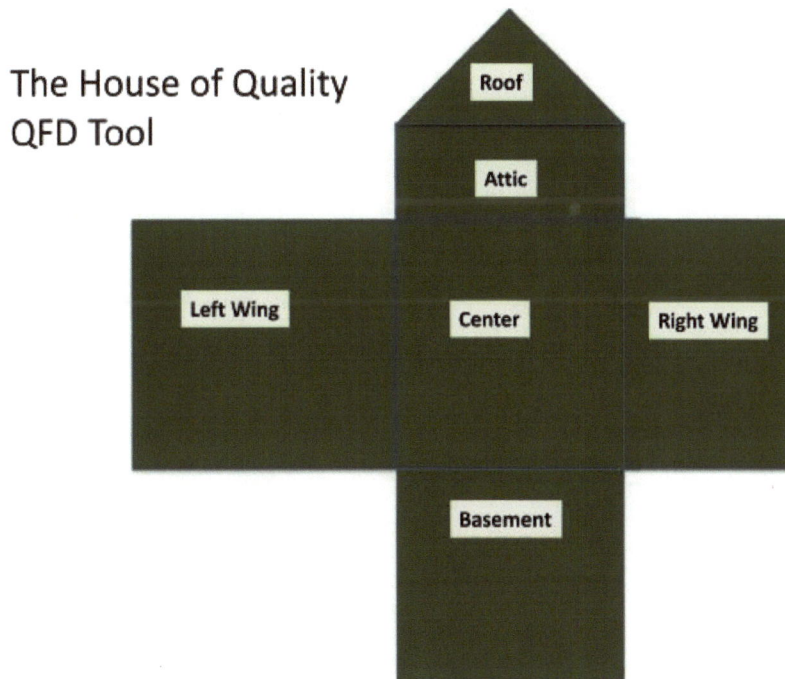

The House of Quality
QFD Tool

Roof

Attic

Left Wing

Center

Right Wing

Basement

10

Figure 13.2 House of Quality

To begin to fill in the House of Quality, we will assemble a design team. The design team should be cross-functional and include representatives from engineering, R&D, production, sales, marketing, customer service, distribution, and purchasing.

As shown in Figure 13.3, the team writes down the customer requirements that it has gleaned from focus groups, surveys, feedback forms, and so on, on the left wing. An importance column to weight each of the customer requirements can also be added. (Wortman, 2004)

If the team has access to the information, it can record customer assessments of the competition's existing products or services and put these on the right wing.

The team then writes product or service characteristics in the attic of the house. The roof will contain a correlation matrix between product characteristics. The team then identifies relationships between customer requirements and product /service characteristics in the center of the house.

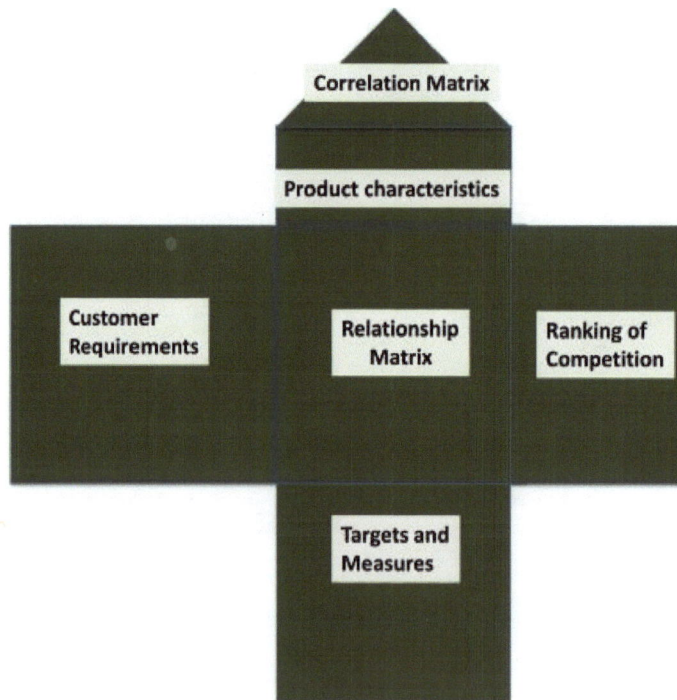

Figure 13.3 Parts of the House or Quality

In the basement, we use a row identifying the measurement unit for each product or service characteristic by recording the performance of existing products or services and determining targets for the measurements.

The House of Quality contains a great deal of information captured in one place. Accordingly, the House of Quality can take days to fully flesh out, even if all the input information is available. While the tool can lead to useful insights, it is not used as often as it could be because filling out the matrices is time consuming and complicated.

13.3   Matrix Diagrams

We can think of the House of Quality as an amalgam of various **matrix diagrams**. Matrix diagrams are used to compare two or more groups at once. Many of these diagrams are named using the letters of the alphabet that they resemble. For example, an L matrix takes the shape shown in Figure 13.4. Note that the grayed-in area forms a sideways L shape.

Group 1	Group 2		
	X	Y	Z
A			
B			
C			
D			

Figure 13.4 Matrix diagram

The L matrix is used to relate two groups to each other and is the most used matrix shape. The relationship matrix of the HOQ, in which customer requirements are compared to product characteristics, is an example of an L-shaped matrix. The prioritization matrix, and stakeholder analysis matrix also make use of this shape.

A roof-shaped matrix relates one group to itself. For example, the roof of the House of Quality shows correlations within the group of product characteristics.

There are other matrix shapes as well. The T matrix show the relationship among three groups in which two group, say B and C, are related to a third group A, but not to each other. The Y matrix relates three groups to each other, comparing them two at a time. The C matrix compares

three groups to each other simultaneously in a 3-dimensional matrix. The X matrix displays relationship among 4 groups, in which each pair of groups is compared in turn. See Tague (2005) for more details.

13.4   Failure Mode & Effects Analysis

Finally, we will learn about Failure Mode and Effects Analysis, or FMEA, and its role in designing a new product or process. FMEA can be used to identify ways that an existing or new process product or service can fail. We estimate the risk associated with each specific failure and attempt to reduce this risk.

We use a cross-functional team once again to create a very detailed FMEA analysis. In evaluating a process, the team lists each process step. For a product or service, the team lists all the features in the left most column of the FMEA matrix. An FMEA template is provided in the course companion spreadsheet and shown in Figure 13.5.

For each process step, the team identifies ways that an error can occur, or for a product, ways that a feature could fail. There can be many, many potential failure modes. For each mode, the team lists out all the effects of the failure. There can be multiple effects for each failure mode. Accordingly, the FMEA table gets very large very quickly. For each of the potential failure modes, the team assigns a *severity* score (S) from 1 to 10, with 10 being the most severe.

For each failure mode, the team also lists all the potential causes of that failure. For each of these, the probability of *occurrence* (O) is scored, with a 1 being extremely low and a 10 being very certain.

FMEA Table

Process Step	Potential Failure Mode	Potential Failure Effects	S e v e r i t y	Potential Causes	O c c u r r e n c e	Current Controls	D e t e c t i o n	R P N
What is the process step and input under investigation?	In what ways does the key input go wrong?	What is the impact on the key output variables?		What causes the key input to go wrong?		What are the existing controls and procedures that prevent the cause of the failure mode?		

Severity, Occurrence and Detection are all rated on a 1-10 scale.
Severity: 10 is the most severe
Occurrence is the likelihood of a failure, 10 is the most likely
Detection: 10 is least noticeable

RPN = Risk Priority Number = S × O × D

17

Figure 13.5 FMEA table

For each potential cause, the team will then list all the current controls in the place that will allow the failure to be detected. For each control, the team will score a *detection* score (D), where 1 is easily detectable and 10 is the least noticeable.

Once the matrix has been completed, a risk priority number (RPN) is calculated for each separate line in the table. RPN is a combination of the severity, occurrence and detection scores, calculated as S x O x D. The higher the RPN, the higher the risk.

The items in the matrix are then sorted in descending order of RPNs, with the second sort key being severity. The team can then identify the riskiest failure modes and try to reduce the RPN of these top items by increasing detection, reducing the probability of occurrence, or eliminating the causes. For items with equal RPNs, the team use the severity rating to rate the tied items.

13.5   Process Decision Program Chart

A **process decision program chart** (PDPC) is used when planning a project, especially if the project is complex, and the cost of failure is high. This tool can be used in designing a new product or used in conjunction with a stakeholder analysis to create a game plan for project success. (Tague, 2005)

The PDPC uses a tree diagram as its foundation. The high-level project activities (or steps of the plan) are laid out on the first level of the tree, followed by main activities and more detailed steps on the second and third levels, respectively. Next, the project leader and team address each detailed step in the third level and brainstorm what could go wrong. These potential problems are written below the step in a fourth level of the tree. The team continues until all project steps have been addressed.

Next, the team goes back and eliminates all brainstormed problems that are improbable or that have very little negative impact on the project. For the remaining problems, a countermeasure is developed by the team. The countermeasure could take the form of a method of eliminating the occurrence of the problem; or a method to mitigate the effect of the problem; or a way to reduce the likelihood of occurrence of a problem. These countermeasures are written as a fifth level of the tree and circled.

Finally, the team evaluates each countermeasure, assessing how practical it is in terms of cost, resources, or time. Impractical countermeasures are marked with an X and ones that seem to be of use are marked with a check.

Once the tree has been cleaned up, the team is left with a plan to address problems they might encounter at each step of the project.

Chapter 14 Closing the Project

Once the Improve phase is complete, and the solutions to the process problem have been implemented, there is a real temptation for the team to disband and move on. The team leader must keep the team focused on completing the last stage of the project, the Control phase. The outcomes of this phase will help ensure that the improvements implemented stick, and that the process owners have the tools to sustain the gains achieved. In addition, the Control phase allows the team to perform a postmortem of the project and share lessons learned with the rest of the organization. Finally, there is a celebration of the achievements of the team.

14.1 Control Plans

A control plan is a document that specifies how the improved process will be monitored going forward to assure that the gains are sustained.

Recall that the process metrics we track need to be linked to customer requirements. Organizations gather Voice of the Customer information, and then translate this feedback into Critical to Quality characteristics (CTQs). These CTQs are statements of what the customer expects or deserves from the organization's products. The CTQs are then linked to key process output variables, or KPOVs. These output variables must be within certain specifications to satisfy the CTQs. Through the Measure and Analyze phases of the Six Sigma project, the team discovers the relationship that links key process input variables (KPIV) to the process outputs (the KPOVs). The equation $Y = f(x)$ represents this relationship between inputs (the x's) that are put through a function, or a process, that lead to outputs or Ys. (Figure 14.1)

Given a choice, it is better to track and monitor input variables. By keeping the input variables within specifications, we can be assured that the outputs will satisfy the CTQs. By focusing on measuring the inputs, we are tracking leading indicators. If an input is out of specification, adjustment can be made before labor and material costs are incurred in the process steps.

Use CTQ Requirements to Develop KPOVs

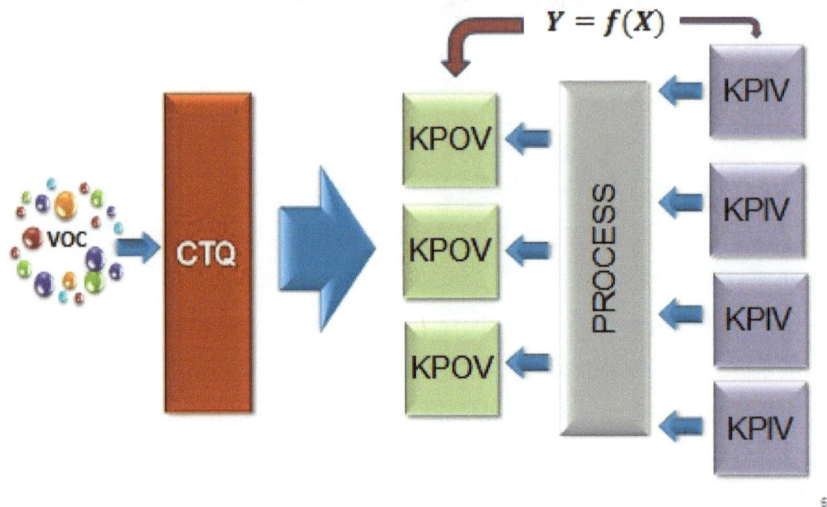

Figure 14.1 CTQ requirements

The complete control plan will specify several items:

1) A summary of process changes
   a. A new standard operating procedure that includes process changes and a visual representation of new work instructions
   b. A SIPOC diagram that gives a 30,000-foot overview of the improve process
2) A listing of key process input and output variables with their specifications
3) A sampling plan that details:
   a. the frequency of sampling
   b. sample size
   c. the variables measured, and their specifications
   d. the measurement equipment used for each metric
   e. party responsible for collecting samples, measuring, tracking, and monitoring the metrics
   f. method for monitoring measurements (control charts, dashboard, etc.)

4) A corrective action plan that details:

    a.   the action taken when a metric signals a change

    b.   a listing of emergency fixes

    c.   identification of person or department responsible for action

5) An audit plan that details:

    a.   the frequency of the audit

    b.   the audit item checklist

    c.   person or entity responsible for performing the audit

    d.   person or department responsible for fixing deficiencies found in the audit

6) Plans for continuous improvement

14.2 Validating the Gains

A major component of any Six Sigma project is its impact on the bottom line, whether the project brings cost savings, increased revenue, or increased profit. Black Belts learn how to calculate sophisticated financial metrics such as return on investment to justify capital expenditures and estimate the value of cost savings. These results are used to project the fiscal impact of the project and are published in the project charter.

During the Control phase, the team validates the projected financial gains. After making process improvements, for example, the costs of scrap and rework can be tracked and compared to the pre-project levels. In addition, the value of increased sales or profit can be captured.

Some financial gains may take longer to validate if, say, there is a long sales cycle. In this case, the control plan may include metrics that measure costs and income throughout the period so that the data can be assessed later.

It must be stressed that the financial gains from a Six Sigma project should capture real dollar savings and income. Including only cost avoidance or using inflated savings projections may produce a favorable financial picture in the short run, but it will eventually degrade the integrity of the Six Sigma program.

14.3 Lessons Learned

The Lessons Learned document is a key outcome of the Control phase. At the team's final meeting, the leader conducts a brainstorming session and debrief to identify what activities, decisions and tools worked well in the project, and which did not. Working separately, the project leader can generate a listing of positive and negatives in terms of how the group worked

together, how meetings were held, and how decisions were made. This document can be published on a shared drive to help future teams avoid the same mistakes and leverage the positive aspects identified by the teams.

14.4 Recognition and Celebration

The team has worked two to nine months on a project, in addition to their "real" jobs. They have coalesced into a highly functioning team and have learned many new skills. In addition, the project has improved a process, increased customer satisfaction, and added to the bottom line. Some recognition and celebration are in order!

The Black Belt will work with the executive steering committee and the project sponsor to identify an appropriate way to recognize the team and celebrate its success. Methods might include a luncheon with upper management, recognition at a company-wide meeting, an article in the company newsletter, or a video featuring team members showing before and after results.

Appendix A Mirasol Hotel Case Study

MIRASOL RESORT

A Lean Six Sigma Case Study

© 2020 University Training Partners

JONATHAN SAND

President Sand Island Resorts

This case is based on a small beach resort hotel recently acquired by Jonathan Sand, President of Sand Island Resorts, LLP. Sand began working in the hotel industry as a teenager, manning the front desk and taking reservations at the local hotel owned by his uncle.

After earning his college degree in manufacturing management, he worked as an assembly line supervisor at an automotive plant as well as a production engineer in the pressed metal shop. When the assembly plant was shut down, Jonathan returned to the East Coast and found a position as a maintenance supervisor at Mirasol, a local beach resort. He found that his industrial background gave him a unique edge in attacking and solving problems in the hospitality sector.

In the past five years, he has held wide-ranging positions at the resort to include accounting supervisor, banquet manager, and housekeeping manager. In 2008, Sand was promoted to general manager of Mirasol, which was then owned by Allendale Holdings, a company that managed seven high-end resorts in South Carolina and Florida.

VISIT SOUTH CAROLINA

Takeover from Allendale Holdings

The economic downturn and credit crunch forced Allendale to liquidate some of its properties at deeply discounted prices in order to remain solvent. When it was announced that Marisol was being sold off, Jonathan contacted his uncle, now retired, and formed the Sand Island Resorts partnership. Sand and his Uncle Fred were able to acquire the property a year ago. Since then, Sand has been enthusiastically making changes to put his own stamp on the resort.

In 2009, the resort was featured in the Condé Nast Traveler magazine, which described it as "a boutique resort that is the ultimate in a relaxing getaway; private and personalized."

"A BOUTIQUE RESORT THAT IS THE ULTIMATE IN A RELAXING GETAWAY; PRIVATE AND PERSONALIZED."

Inspired by the Condé Nast review, Sand crafted the new company mission statement and has it displayed in framed posters that hang in the main lobby and the back offices:

"Sand Island Resorts is committed to providing its member guests with personalized service and a relaxing resort experience that consistently delivers excellent quality and high value."

FACILITIES

The Mirasol resort is located on a small barrier island along the Intercostal Waterway, accessible only by passenger ferry. The ferry, operated by the State, makes six round trips each day. Automobiles are restricted on the island, and the main means of transportation are golf carts and bicycles.

The small resort has 50 suite-style rooms in the two-story main building. There are two pools, one with a swim-up bar, as well as three heated spas on the property. Water sports are accessible from the marina located to the west of the property, reachable by a complimentary golf cart shuttle. The beach is accessible via a boardwalk, and guests can reserve cabanas, umbrellas and chairs, surf boards, boogie boards and snorkel equipment at the beach pavilion.

There is one restaurant serving breakfast and dinner daily in the main building, a small bar located off of the main lobby, a poolside snack bar, and a separate banquet/meeting facility that can accommodate up to 175 guests. A gift shop is in the lobby, selling sundry items and resort-branded tote bags, beach towels, golf shirts and Tervis tumblers. A separate exercise and spa facility is located next to the pool area. The gym has four stationary bikes, three treadmills, a Stairmaster, an aerobics/yoga room and locker rooms with showers. The spa has two treatment rooms, a manicure and pedicure station and 2 outside massage cabanas.

The facilities building is located at the north end of the property, and contains a commercial laundry facility, a maintenance garage for the resort golf carts and bicycles, a tool shop, two storage rooms, a hurricane preparedness room and the engineering department offices.

Appendix A Mirasol Hotel Case Study © 2021 University Training Partners

ECONOMIC & MARKET PRESSURES

Allendale Holdings saw a 15% decline in room occupancy and special events bookings at the start of the economic downturn in 2007. Since purchasing the resort, Sand Island Resorts has experienced a 2-3% growth in these rates.

Sand attributes the growth to ads he ran in newspapers in the Northeast, and would like to expand the advertising campaign. His uncle would rather concentrate advertising efforts on adjoining states due to the cost of gasoline. Reina Andersen, Marketing Director, has proposed partnering with online booking sites to offer special accommodation packages at deeply discounted rates. Sand is reluctant to run the special rates because he is afraid of diluting the Mirasol brand.

In a recent meeting, the Director of Recreation and Spa Services, Linda Petrucelli, suggested offering a spa destination package during the "shoulder season," complete with spa cuisine, spa treatments and weeklong courses taught by celebrity yoga masters. Sand thinks that demand for such event is not strong enough to offset the cost.

While Allendale was experiencing a cash flow shortage, it postponed routine maintenance and upkeep. Upon acquiring the property, SIR had to perform costly repairs to the pool deck, air conditioning system and septic system.

In addition, SIR repainted the exterior, made improvements in the restaurant and spa facilities, and purchased new furniture for the beach cottages.

Over the past year, the cost of maintenance, repairs and improvements have eaten into SIR profits. Controller Greg Schultz has suggested reducing the number of resort amenities to offset increased expenses, but Sand is worried that fewer services will translate into fewer bookings.

In addition to economic worries, Sand is concerned that a new hotel recently opened on the south end of the island will dilute his market share. The hotel is an all-inclusive resort owned by a national chain that caters to families with children. He has a suspicion that that his more price sensitive customers are flocking to the new hotel because it offers more activities and lower prices. In addition, couples seeking a destination wedding venue may be enticed by the larger banquet facility. This would be a huge blow to Mirasol's revenue, since weddings make up 20% of its bookings from April through October.

Appointment bookings at the spa have dropped, making it difficult for Sand to justify keeping his three aestheticians on staff. In previous years, the spa was a profit center for the resort. Sand feels that the spa is disorganized and suffers from a lack of consistency in atmosphere: depending who is at the desk, the customer experience could range from muted and distant to boisterously friendly.

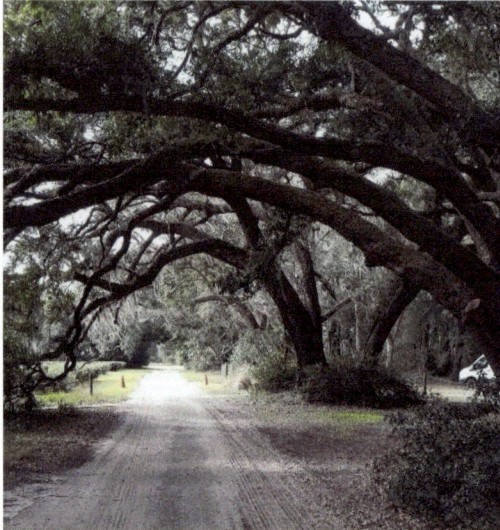

Reina Andersen has been pushing to expand the merchandising business by marketing a line of Mirasol stemware and bar accessories to sell through department stores such as Dilliard's and Nordstrom. She feels this will create new interest for the target market and will be a natural addition to the wedding gift registries of couples married at Mirasol. Both Sand and Greg Shultz, the Controller, see potential in the idea, but are reluctant to make the capital investment for the new line given the current economic climate.

OPERATIONAL ISSUES

According to Eli Guzmán, the Housekeeping Manager, water usage in the on-premises laundry facility has increased by 25% compared to last year. Over the past few months, he has worked with the service contractor to adjust the wash cycle parameters, but the changes have not reduced water usage to target levels. Guzmán has suggested switching to a contractor that leases wash water recycling systems, but Greg Schultz has rejected the idea based on the high installation fee and monthly leasing costs.

The golf cart maintenance garage has experienced parts shortages in the past six months due supplier problems in Japan. This past summer, three of the ten large, 8-passenger golf carts were out of commission, which caused delays in shuttling guests to and from the ferry dock. Chief Engineer Jim Sutton would like to stockpile replacement parts to avoid shortages in the future, but the garage space is already too small for the number of vehicles they service, and parts and tools are often difficult to find.

Inclement weather or equipment problems can cause delays in the ferry schedule, which can leave Mirasol understaffed at crucial times. Sand has adopted an "all hands on deck" approach when these incidents occur, but knows that the quality and speed of the work performed suffers. Last Friday, he and his uncle cleaned 4 rooms when the ferry was two hours late due to engine trouble. Later, his housekeeping staff told him that he had missed vacuuming under the beds, failed to replenish the Tazo tea bags, and did not fold the duvet according to standard.

There has been high absenteeism and significant turnover in the maintenance and housekeeping departments in the past 6 months. Sand believes that the new resort may be offering higher pay. Training new staff members has taken time away from other duties of the Chief Engineer and Housekeeping Manager.

After the sale of Mirasol, the Executive Chef and Banquet Manger decided to remain with Allendale Holdings and were reassigned to a resort on the Gulf Coast of Florida. Since then, Frank Soule, the sous chef, has taken over the Executive Chef and Banquet Manager duties while an outside firm conducts a talent search. Frank has handled his new responsibilities reasonably well, but Sand fears that he may quit due to exhaustion if new staff members are not found quickly.

The quality of service has suffered recently. In the past, an average of 4 complaint cards per month was received by the front desk. In June, 25 complaints were filed by guests. Customers have complained about lengthy waits for the shuttle, special requests that were ignored, incorrect room service orders, inoperable air conditioners and overcharges upon check out. The overcharges have particularly grated on Sand, who wants to make sure this mistake never happens again.

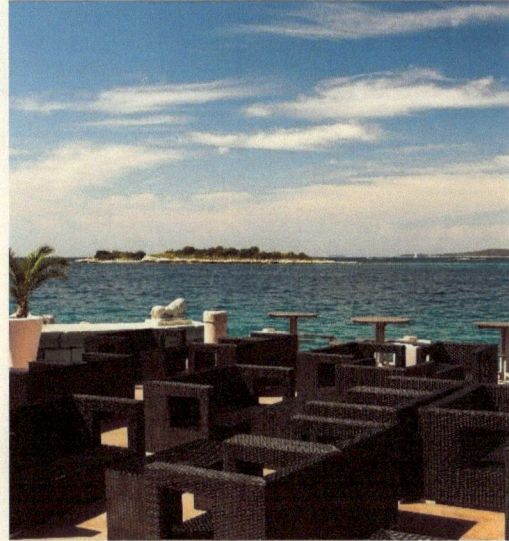

SIX SIGMA IMPLEMENTATION

Sand was involved in quality initiatives at the automotive plant where he previously worked, and had met with the plant's quality manager to discuss data collection and SPC charts a few times. He had tried to incorporate some quality principles into the business operations while he was general manager of the resort back in the Allendale days, but he had neither the experience nor the corporate support to implement a full-fledged program. While attending a hotel and hospitality convention last fall, Sand heard a presentation from an owner of a mid-sized hotel that had decreased costs and increased occupancy by implementing Six Sigma and Lean techniques. Sand began researching Six Sigma and decided to use the DMAIC process to address some of his resort's problems. He himself became Green Belt certified, along with the housekeeping manager and chief engineer.

The other directors and managers have gone through a 2-day Yellow Belt course, led by an outside consultant who is a Six Sigma Black Belt. With the Black Belt's guidance, SIR plans to identify and complete 3-4 Six Sigma projects this year.

Data Tables prepared by Greg Schultz, Reina Anderson and Dave Harris

All data from previous year

Occupancy rate by month

January	10%
February	15%
March	20%
April	45%
May	50%
June	75%
July	70%
August	65%
September	65%
October	45%
November	20%
December	30%

Average Length of Stay in days by month

January	1.6
February	1.7
March	2.8
April	3.6
May	2.5
June	3.5
July	3.9
August	3.6
September	3.7
October	1.7
November	1.5
December	2.6

Revenue per Room by month

January	$	1,395
February	$	1,890
March	$	3,255
April	$	7,628
May	$	9,300
June	$	17,438
July	$	17,360
August	$	16,120
September	$	12,188
October	$	7,324
November	$	2,400
December	$	4,883

Percent Bookings by reservation channel

Hotel website	54%
Direct Calls	16%
Online booking agencies	30%

Total Room Changeovers by Month

January	78
February	99
March	89
April	150
May	248
June	257
July	223
August	224
September	211
October	328
November	160
December	143

Note: A changeover occurs when a guest checks out of a room and a new guest checks in

Total complaints by month

January	2
February	2
March	4
April	4
May	5
June	25
July	20
August	24
September	18
October	27
November	15
December	10

Percent complaints per month by room changeover

January	3%
February	2%
March	4%
April	3%
May	2%
June	10%
July	9%
August	11%
September	9%
October	8%
November	9%
December	7%

Sand Island Resorts - Mirasol Hotel Organizational Chart

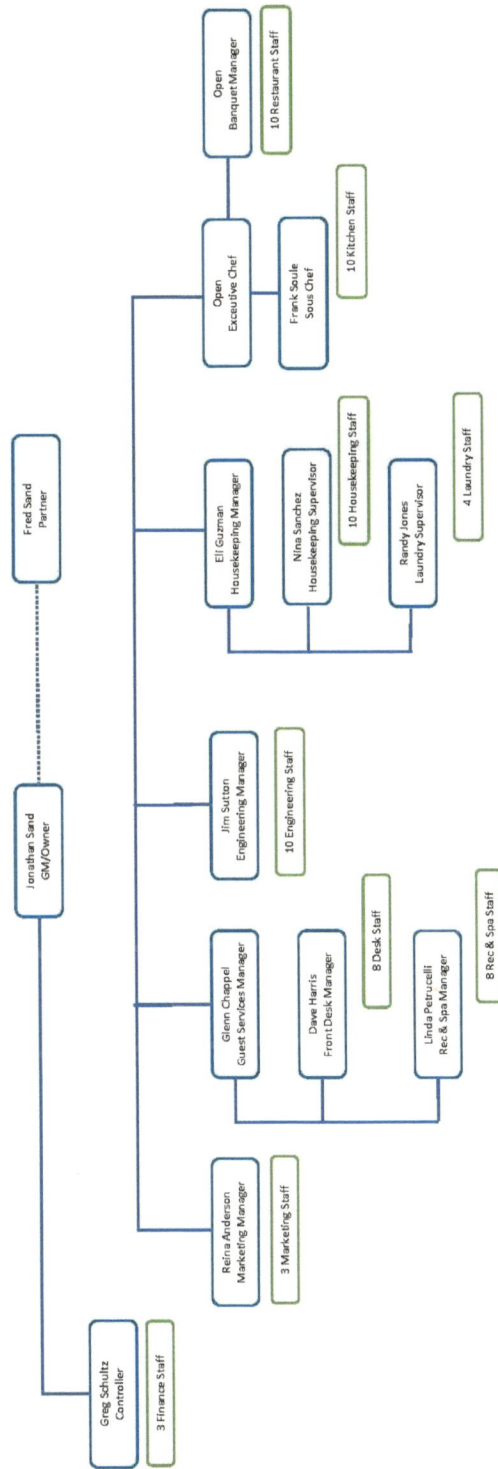

Fred Sand
Partner

Jonathan Sand
GM/Owner

Greg Schultz
Controller
- 3 Finance Staff

Reina Anderson
Marketing Manager
- 3 Marketing Staff

Glenn Chappel
Guest Services Manager

Dave Harris
Front Desk Manager
- 8 Desk Staff

Linda Petrucelli
Rec & Spa Manager
- 8 Rec & Spa Staff

Jim Sutton
Engineering Manager
- 10 Engineering Staff

Eli Guzman
Housekeeping Manager

Nina Sanchez
Housekeeping Supervisor
- 10 Housekeeping Staff

Randy Jones
Laundry Supervisor
- 4 Laundry Staff

Open
Executive Chef

Frank Soule
Sous Chef
- 10 Kitchen Staff

Open
Banquet Manager
- 10 Restaurant Staff

Appendix A Mirasol Hotel Case Study © 2021 University Training Partners

Appendix B The ASQ Black Belt Body of Knowledge[2]

---

[2] See http://asq.org/cert/six-sigma-black-belt/bok

AMERICAN SOCIETY FOR QUALITY SIX SIGMA BLACK BELT BODY OF KNOWLEDGE

The topics in this Body of Knowledge include additional detail in the form of subtext explanations and the cognitive level at which the questions will be written. This information will provide useful guidance for both the Examination Development Committee and the candidates preparing to take the exam. The subtext is not intended to limit the subject matter or be all-inclusive of what might be covered in an exam. It is meant to clarify the type of content to be included in the exam. The descriptor in parentheses at the end of each entry refers to the maximum cognitive level at which the topic will be tested. A more complete description of cognitive levels is provided at the end of this document.

I. Enterprise-Wide Deployment [9 Questions]

    A. Enterprise-wide view

        1. History of continuous improvement Describe the origins of continuous improvement and its impact on other improvement models. (Remember)

        2. Value and foundations of Six Sigma Describe the value of Six Sigma, its philosophy, history, and goals. (Understand)

        3. Value and foundations of Lean Describe the value of Lean, its philosophy, history, and goals. (Understand)

        4. Integration of Lean and Six Sigma Describe the relationship between Lean and Six Sigma. (Understand)

        5. Business processes and systems Describe the relationship among various business processes (design, production, purchasing, accounting, sales, etc.) and the impact these relationships can have on business systems. (Understand)

        6. Six Sigma and Lean applications Describe how these tools are applied to processes in all types of enterprises: manufacturing, service, transactional, product and process design, innovation, etc. (Understand)

    B. Leadership

        1. Enterprise leadership responsibilities Describe the responsibilities of executive leaders and how they affect the deployment of Six Sigma in terms

of providing resources, managing change, communicating ideas, etc. (Understand)

2. Organizational roadblocks Describe the impact an organization's culture and inherent structure can have on the success of Six Sigma, and how deployment failure can result from the lack of resources, management support, etc.; identify and apply various techniques to overcome these barriers. (Apply)

3. Change management Describe and use various techniques for facilitating and managing organizational change. (Apply)

   1. Six Sigma projects and kaizen events Describe how projects and kaizen events are selected, when to use Six Sigma instead of other problem-solving approaches, and the importance of aligning their objectives with organizational goals. (Apply)

   2. Six Sigma roles and responsibilities Describe the roles and responsibilities of Six Sigma participants: black belt, master black belt, green belt, champion, process owners and project sponsors. (Understand)

II. Organizational Process Management and Measures [9 Questions]

   A. Impact on stakeholders Describe the impact Six Sigma projects can have on customers, suppliers, and other stakeholders. (Understand)

   B. Critical to x (CTx) requirements Define and describe various CTx requirements (critical to quality (CTQ), cost (CTC), process (CTP), safety (CTS), delivery (CTD), etc.) and the importance of aligning projects with those requirements. (Apply)

   C. Benchmarking Define and distinguish between various types of benchmarking, including best practices, competitive, collaborative, etc. (Apply)

   D. Business performance measures Define and describe various business performance measures, including balanced scorecard, key performance indicators (KPIs), the financial impact of customer loyalty, etc. (Understand)

   E. Financial measures Define and use financial measures, including revenue growth, market share, margin, cost of quality (COQ), net present value (NPV), return on investment (ROI), cost- benefit analysis, etc. (Apply)

III. Team Management [16 Questions]

    A.   Team formation

        1. Team types and constraints Define and describe various types of teams (e.g., formal, informal, virtual, cross- functional, self-directed, etc.), and determine what team model will work best for a given situation. Identify constraining factors including geography, technology, schedules, etc. (Apply)

        2. Team roles Define and describe various team roles and responsibilities, including leader, facilitator, coach, individual member, etc. (Understand)

        3. Team member selection Define and describe various factors that influence the selection of team members, including required skills sets, subject matter expertise, availability, etc. (Apply)

        4. Launching teams Identify and describe the elements required for launching a team, including having management support, establishing clear goals, ground rules and timelines, and how these elements can affect the team's success. (Apply)

    B.   Team facilitation

        1. Team motivation Describe and apply techniques that motivate team members and support and sustain their participation and commitment. (Apply)

        2. Team stages Facilitate the team through the classic stages of development: forming, storming, norming, performing and adjourning. (Apply)

        3. Team communication Identify and use appropriate communication methods (both within the team and from the team to various stakeholders) to report progress, conduct milestone reviews and support the overall success of the project. (Apply)

    C.   Team dynamics Identify and use various techniques (e.g., coaching, mentoring, intervention, etc.) to overcome various group dynamic challenges, including overbearing/dominant or reluctant participants, feuding and other forms of unproductive disagreement, unquestioned acceptance of opinions as facts, groupthink, floundering, rushing to accomplish or finish, digressions, tangents, etc. (Evaluate)

    Appendix B The ASQ Black Belt Body of Knowledge © 2021 University Training Partners

D. Time management for teams Select and use various time management techniques including publishing agendas with time limits on each entry, adhering to the agenda, requiring pre-work by attendees, ensuring that the right people and resources are available, etc. (Apply)

E. Team decision-making tools Define, select, and use tools such as brainstorming, nominal group technique, multi-voting, etc. (Apply)

F. Management and planning tools Define, select, and apply the following tools: affinity diagrams, tree diagrams, process decision program charts (PDPC), matrix diagrams, interrelationship digraphs, prioritization matrices and activity network diagrams. (Apply)

G. Team performance evaluation and reward Measure team progress in relation to goals, objectives and other metrics that support team success and reward and recognize the team for its accomplishments. (Analyze)

IV. Define [15 Questions]

    A. Voice of the customer

        1. Customer identification Segment customers for each project and show how the project will impact both internal and external customers. (Apply)

        2. Customer feedback Identify and select the appropriate data collection method (surveys, focus groups, interviews, observation, etc.) to gather customer feedback to better understand customer needs, expectations, and requirements. Ensure that the instruments used are reviewed for validity and reliability to avoid introducing bias or ambiguity in the responses. (Apply)

        3. Customer requirements Define, select, and use appropriate tools to determine customer requirements, such as CTQ flow-down, quality function deployment (QFD) and the Kano model. (Apply)

    B. Project charter

        1. Problem statement Develop and evaluate the problem statement in relation to the project's baseline performance and improvement goals. (Create)

        2. Project scope Develop and review project boundaries to ensure that the project has value to the customer. (Analyze)

        3. Goals and objectives Develop the goals and objectives for the project on the basis of the problem statement and scope. (Apply)

4. Project performance measures Identify and evaluate performance measurements (e.g., cost, revenue, schedule, etc.) that connect critical elements of the process to key outputs. (Analyze)

C. Project tracking Identify, develop, and use project management tools, such as schedules, Gantt charts, toll-gate reviews, etc., to track project progress. (Create)

V. Measure [26 Questions]

A. Process characteristics

1. Input and output variables Identify these process variables and evaluate their relationships using SIPOC and other tools. (Evaluate)

2. Process flow metrics Evaluate process flow and utilization to identify waste and constraints by analyzing work in progress (WIP), work in queue (WIQ), touch time, takt time, cycle time, throughput, etc. (Evaluate)

3. Process analysis tools Analyze processes by developing and using value stream maps, process maps, flowcharts, procedures, work instructions, spaghetti diagrams, circle diagrams, etc. (Analyze)

B. Data collection

1. Types of data Define, classify, and evaluate qualitative and quantitative data, continuous (variables) and discrete (attributes) data and convert attributes data to variables measures when appropriate. (Evaluate)

2. Measurement scales Define and apply nominal, ordinal, interval and ratio measurement scales. (Apply)

3. Sampling methods Define and apply the concepts related to sampling (e.g., representative selection, homogeneity, bias, etc.). Select and use appropriate sampling methods (e.g., random sampling, stratified sampling, systematic sampling, etc.) that ensure the integrity of data. (Evaluate)

4. Collecting data Develop data collection plans, including consideration of how the data will be collected (e.g., check sheets, data coding techniques, automated data collection, etc.) and how it will be used. (Apply)

C. Measurement systems

1. Measurement methods Define and describe measurement methods for both continuous and discrete data. (Understand)

2. Measurement systems analysis Use various analytical methods (e.g., repeatability and reproducibility (R&R), correlation, bias, linearity, precision to tolerance, percent agreement, etc.) to analyze and interpret measurement system capability for variables and attributes measurement systems. (Evaluate)

3. Measurement systems in the enterprise Identify how measurement systems can be applied in marketing, sales, engineering, research and development (R&D), supply chain management, customer satisfaction and other functional areas. (Understand)

4. Metrology Define and describe elements of metrology, including calibration systems, traceability to reference standards, the control and integrity of standards and measurement devices, etc. (Understand)

D. Basic statistics

1. Basic terms Define and distinguish between population parameters and sample statistics (e.g., proportion, mean, standard deviation, etc.) (Apply)

2. Central limit theorem Describe and use this theorem and apply the sampling distribution of the mean to inferential statistics for confidence intervals, control charts, etc. (Apply)

3. Descriptive statistics Calculate and interpret measures of dispersion and central tendency and construct and interpret frequency distributions and cumulative frequency distributions. (Evaluate)

4. Graphical methods Construct and interpret diagrams and charts, including box-and-whisker plots, run charts, scatter diagrams, histograms, normal probability plots, etc. (Evaluate)

5. Valid statistical conclusions Define and distinguish between enumerative (descriptive) and analytic (inferential) statistical studies and evaluate the results to draw valid conclusions. (Evaluate)

E. Probability

1. Basic concepts Describe and apply probability concepts such as independence, mutually exclusive events, multiplication rules, complementary probability, joint occurrence of events, etc. (Apply)

2. Commonly used distributions Describe, apply, and interpret the following distributions: normal, Poisson, binomial, chi square, Student's t, and F distributions. (Evaluate)

3. Other distributions Describe when and how to use the following distributions: hypergeometric, bivariate, exponential, lognormal and Weibull. (Apply)

F. Process capability

1. Process capability indices Define, select, and calculate C_p and C_{pk} to assess process capability. (Evaluate)

2. Process performance indices Define, select, and calculate P_p, P_{pk} and C_{pm} to assess process performance. (Evaluate)

3. Short-term and long-term capability Describe and use appropriate assumptions and conventions when only short-term data or attributes data are available and when long-term data are available. Interpret the relationship between long-term and short-term capability. (Evaluate)

4. Process capability for non-normal data Identify non-normal data and determine when it is appropriate to use Box-Cox or other transformation techniques. (Apply)

5. Process capability for attributes data Calculate the process capability and process sigma level for attributes data. (Apply)

6. Process capability studies Describe and apply elements of designing and conducting process capability studies, including identifying characteristics and specifications, developing sampling plans and verifying stability and normality. (Evaluate)

7. Process performance vs. specification Distinguish between natural process limits and specification limits, and calculate process performance metrics such as percent defective, parts per million (PPM), defects per million opportunities (DPMO), defects per unit (DPU), process sigma, rolled throughput yield (RTY), etc. (Evaluate)

VI. Analyze [24 Questions]

A. Measuring and modeling relationships between variables

1. Correlation coefficient Calculate and interpret the correlation coefficient and its confidence interval and describe the difference between correlation and causation. (Analyze) NOTE: Serial correlation will not be tested.

2. Regression Calculate and interpret regression analysis and apply and interpret hypothesis tests for regression statistics. Use the regression model for estimation and prediction, analyze the uncertainty in the estimate, and perform a residuals analysis to validate the model. (Evaluate) NOTE: Models that have non-linear parameters will not be tested.

3. Multivariate tools Use and interpret multivariate tools such as principal components, factor analysis, discriminant analysis, multiple analysis of variance (MANOVA), etc., to investigate sources of variation. (Analyze)

4. Multi-vari studies Use and interpret charts of these studies and determine the difference between positional, cyclical, and temporal variation. (Analyze)

5. Attributes data analysis Analyze attributes data using logit, probit, logistic regression, etc., to investigate sources of variation. (Analyze)

B. Hypothesis testing

1. Terminology Define and interpret the significance level, power, type I, and type II errors of statistical tests. (Evaluate)

2. Statistical vs. practical significance Define, compare, and interpret statistical and practical significance. (Evaluate)

3. Sample size Calculate sample size for common hypothesis tests (e.g., equality of means, equality of proportions, etc.). (Apply)

4. Point and interval estimates Define and distinguish between confidence and prediction intervals. Define and interpret the efficiency and bias of estimators. Calculate tolerance and confidence intervals. (Evaluate)

5. Tests for means, variances and proportions Use and interpret the results of hypothesis tests for means, variances and proportions. (Evaluate)

6. Analysis of variance (ANOVA) Select, calculate, and interpret the results of ANOVAs. (Evaluate)

7. Goodness-of-fit (chi square) tests Define, select, and interpret the results of these tests. (Evaluate)

---

8. Contingency tables Select, develop, and use contingency tables to determine statistical significance. (Evaluate)

9. Non-parametric tests Select, develop, and use various non-parametric tests, including Mood's Median, Levene's test, Kruskal-Wallis, Mann-Whitney, etc. (Evaluate)

C. Failure mode and effects analysis (FMEA) Describe the purpose and elements of FMEA, including risk priority number (RPN), and evaluate FMEA results for processes, products, and services. Distinguish between design FMEA (DFMEA) and process FMEA (PFMEA) and interpret results from each. (Evaluate)

D. Additional analysis methods

1. Gap analysis Use various tools and techniques (gap analysis, scenario planning, etc.) to compare the current and future state in terms of pre-defined metrics. (Analyze)

2. Root cause analysis Define and describe the purpose of root cause analysis, recognize the issues involved in identifying a root cause, and use various tools (e.g., the 5 whys, Pareto charts, fault tree analysis, cause and effect diagrams, etc.) for resolving chronic problems. (Evaluate)

3. Waste analysis Identify and interpret the 7 classic wastes (overproduction, inventory, defects, over-processing, waiting, motion and transportation) and other forms of waste such as resource under-utilization, etc. (Analyze)

VII. Improve [23 Questions]

A. Design of experiments (DOE)

1. Terminology Define basic DOE terms, including independent and dependent variables, factors and levels, response, treatment, error, etc. (Understand)

2. Design principles Define and apply DOE principles, including power and sample size, balance, repetition, replication, order, efficiency, randomization, blocking, interaction, confounding, resolution, etc. (Apply)

3. Planning experiments Plan, organize and evaluate experiments by determining the objective, selecting factors, responses and measurement methods, choosing the appropriate design, etc. (Evaluate)

4. One-factor experiments Design and conduct completely randomized, randomized block and Latin square designs and evaluate their results. (Evaluate)

5. Two-level fractional factorial experiments Design, analyze and interpret these types of experiments and describe how confounding affects their use. (Evaluate)

6. Full factorial experiments Design, conduct and analyze full factorial experiments. (Evaluate)

B. Waste elimination Select and apply tools and techniques for eliminating or preventing waste, including pull systems, kanban, 5S, standard work, poka-yoke, etc. (Analyze)

C. Cycle-time reduction Use various tools and techniques for reducing cycle time, including continuous flow, single-minute exchange of die (SMED), etc. (Analyze)

D. Kaizen and kaizen blitz Define and distinguish between these two methods and apply them in various situations. (Apply)

E. Theory of constraints (TOC) Define and describe this concept and its uses. (Understand)

F. Implementation Develop plans for implementing the improved process (i.e., conduct pilot tests, simulations, etc.), and evaluate results to select the optimum solution. (Evaluate)

G. Risk analysis and mitigation Use tools such as feasibility studies, SWOT analysis (strengths, weaknesses, opportunities, and threats), PEST analysis (political, environmental, social, and technological) and consequential metrics to analyze and mitigate risk. (Apply)

VIII. Control [21 Questions]

A. Statistical process control (SPC)

1. Objectives Define and describe the objectives of SPC, including monitoring and controlling process performance, tracking trends, runs, etc., and reducing variation in a process. (Understand)

2. Selection of variables Identify and select critical characteristics for control chart monitoring. (Apply)

3. Rational subgrouping Define and apply the principle of rational subgrouping. (Apply)

4. Control chart selection Select and use the following control charts in various situations: X-bar – R, $\overline{X}$ – s, individual and moving range (ImR), p, np, c, u, short-run SPC and moving average. (Apply)

5. Control chart analysis Interpret control charts and distinguish between common and special causes using rules for determining statistical control. (Analyze)

B. Other control tools

1. Total productive maintenance (TPM) Define the elements of TPM and describe how it can be used to control the improved process. (Understand)

2. Visual factory Define the elements of a visual factory and describe how they can help control the improved process. (Understand)

C. Maintain controls

1. Measurement system re-analysis Review and evaluate measurement system capability as process capability improves and ensure that measurement capability is sufficient for its intended use. (Evaluate)

2. Control plan Develop a control plan for ensuring the ongoing success of the improved process including the transfer of responsibility from the project team to the process owner. (Apply)

D. Sustain improvements

1. Lessons learned Document the lessons learned from all phases of a project and identify how improvements can be replicated and applied to other processes in the organization. (Apply)

2. Training plan deployment Develop and implement training plans to ensure continued support of the improved process. (Apply)

3. Documentation Develop or modify documents including standard operating procedures (SOPs), work instructions, etc., to ensure that the improvements are sustained over time. (Apply)

4. Ongoing evaluation Identify and apply tools for ongoing evaluation of the improved process, including monitoring for new constraints, additional opportunities for improvement, etc. (Apply)

IX. Design for Six Sigma (DFSS) Frameworks and Methodologies [7 Questions]

    A. Common DFSS methodologies Identify and describe these methodologies. (Understand)

        1. DMADV (define, measure, analyze, design, and validate)

        2. DMADOV (define, measure, analyze, design, optimize and validate)

    B. Design for X (DFX) Describe design constraints, including design for cost, design for manufacturability and producibility, design for test, design for maintainability, etc. (Understand)

    C. Robust design and process Describe the elements of robust product design, tolerance design and statistical tolerancing. (Apply)

    D. Special design tools

        1. Strategic Describe how Porter's five forces analysis, portfolio architecting, and other tools can be used in strategic design and planning. (Understand)

        2. Tactical Describe and use the theory of inventive problem-solving (TRIZ), systematic design, critical parameter management and Pugh analysis in designing products or processes. (Apply)

Levels of Cognition based on Bloom's Taxonomy – Revised (2001)

In addition to **content** specifics, the subtext for each topic in this BOK also indicates the intended **complexity level** of the test questions for that topic. These levels are from "Levels of Cognition" (from Bloom's Taxonomy – Revised, 2001). They are in rank order - from least complex to most complex.

Remember

Recall or recognize terms, definitions, facts, ideas, materials, patterns, sequences, methods, principles, etc.

Understand

Read and understand descriptions, communications, reports, tables, diagrams, directions, regulations, etc.

Apply

Know when and how to use ideas, procedures, methods, formulas, principles, theories, etc.

Analyze

Break down information into its constituent parts and recognize their relationship to one another and how they are organized; identify sublevel factors or salient data from a complex scenario.

Evaluate

Make judgments about the value of proposed ideas, solutions, etc., by comparing the proposal to specific criteria or standards.

Create

Put parts or elements together in such a way as to reveal a pattern or structure not clearly there before; identify which data or information from a complex set is appropriate to examine further or from which supported conclusions can be drawn.

About the Author

Mary McShane-Vaughn is the founder of University Training Partners, a company that develops and delivers Lean Six Sigma and statistical training, both in-person and online. To date, she has trained more than 1100 Yellow, Green and Black Belts.

Her course offerings can be found at https://courses.6sigma.university

Mary earned her Ph.D. in Industrial Engineering and MS in Statistics from the Georgia Institute of Technology and a BS in Industrial Engineering from Kettering University. Dr. McShane-Vaughn was a tenured faculty member at Southern Polytechnic State University. For eight years she directed and grew the Master of Science in Quality Assurance program and taught statistics, statistical quality control, linear regression, and design of experiments in the graduate program.

Before her career in academics, Dr. McShane-Vaughn worked for 15 years in industry as a quality engineer and statistician in the automotive, medical device manufacturing, consumer products testing and revenue management industries.

Mary is co-author of the *Quality Inspector Handbook*, and author of *The Probability Handbook* and *The Probability Workbook*, all published by ASQ Quality Press. She is an editorial review board member for the Lean & Six Sigma Review. Dr. McShane-Vaughn is a senior member of the American Society for Quality and a member of the Institute of Industrial Engineers and the American Statistical Association. She holds ASQ certifications as a Six Sigma Black Belt, Quality Engineer and Reliability Engineer.

Bibliography

Bothe, Davis R., (2002) "Statistical Reason for the 1.5 Sigma Shift," Quality Engineering, 14:3, 479-487.

Bowman, Sharon L. (2005) *The Ten-Minute Trainer*. San Francisco, CA: Pfeiffer, a Wiley Imprint.

Cameron, Gordon. (2010) *Trizics*. Lexington, Kentucky: Create Space

Eckerson, Wayne W. (2006) *Performance Dashboards, Measuring Monitoring and Managing Your Business*. Hoboken, New Jersey: John Wiley & Sons, Inc.

Eckes, George. (2003) *Six Sigma Team Dynamics, The Elusive Key to Project Success*. Hoboken, New Jersey: John Wiley & Sons, Inc.

George, Michael; Maxey, John; Rowlands, David; and Price, Mark. (2004) *The Lean Six Sigma Pocket Toolbook*. New York: McGraw-Hill Education.

Glaveski, Steve. (2019) Harvard Business Review. *Where Companies Go Wrong with Learning and Development*. [online] Available at: https://hbr.org/2019/10/where-companies-go-wrong-with-learning-and-development.

Goldratt, Eliyahu M. (1990) *Theory of Constraints*. Great Barrington, Massachusetts: The North River Press.

Gygi, Craig and Williams, Brian. (2012). *Six Sigma for Dummies, 2nd edition*. Hoboken, New Jersey: John Wiley & Sons, Inc.

Harry, Mikel and Schroeder, Richard (2000), *Six Sigma, The Breakthrough Management Strategy Revolutionizing the World's Top Corporations*, New York: Currency Books, Random House.

Harry, Mikel J. and Spearman, J. Ronald (1992), *Six Sigma Producibility Analysis and Process Characterization,* Reading, Massachusetts: Addison-Wesley Publishing Company.

Hopp, Wallace J. and Spearman, Mark L. (2001). *Factory Physics, 2nd edition*. Boston: McGraw-Hill Higher Education.

Juran, J.M. and Gryna, Frank M, Jr. (1980). *Quality Planning and Analysis, 2nd edition.* New York: McGraw-Hill Publishing Company.

Kubiak, T.M. and Benbow, Donald W. (2009). *The Certified Six Sigma Black Belt Handbook, 2nd edition.* Milwaukee, Wisconsin: ASQ Quality Press.

Lencioni, Patrick. (2012). *The Five Dysfunctions of a Team, 2nd edition.* San Francisco: Pfeiffer.

Liker, Jeffrey K. (2004). *The Toyota Way, 14 Management Principles from the World's Greatest Manufacturer.* New York: McGraw-Hill.

Little, John D.C. and Graves, Stephen C. (2008). Chapter 5 Little's Law, *Building Intuition: Insights from Basic Operations Management Models and Principles,* D. Chhajed and T.J. Lowe(eds.). Springer Science + Business Media, LLC.

Martin, Karen and Osterling, Mike. (2007). *The Kaizen Event Planner, Achieving Rapid Improvement in Office, Service, and Technical Environments.* Boca Raton, Florida: CRC Press.

McCarty, Thomas; Daniels, Lorraine; Bremer, Michael; and Gupta, Praveen. (2005). *The Six Sigma Black Belt Handbook.* New York: McGraw-Hill.

Nash, Mark A. and Poling, Sheila R. (2008). *Mapping the Total Value Stream, A Comprehensive Guide for Production and Transactional Processing.* New York: CRC Press.

Okes, Duke. (2009). *Root Cause Analysis, The Core of Problem Solving and Corrective Action.* Milwaukee, Wisconsin: ASQ Quality Press.

Owens, Tracy L. (2012). *Six Sigma Green Belt, Round 2.* Milwaukee, Wisconsin: ASQ Quality Press.

Pande, Peter S., Neuman, Robert P., Cavanagh, Roland R. (2002). *The Six Sigma Way Team Fieldbook.* McGraw-Hill.

Phillips, Lawrence (2011) http://www.slideshare.net/LawrencePhillips/kano-model-rev-1

Pike, Robert W. (2003). *Creative Training Techniques Handbook.* Amherst, Massachusetts: HRD Press, Inc.

Rother, Mike and Shook, John. (1999). *Learning to See*. Cambridge, Massachusetts: Lean Enterprise Institute.

Shingo, Shigeo. (1986). *Zero Quality Control: Source Inspection and the Poka-yoke System*. Cambridge, Massachusetts: Productivity Press

Shingo, Shigeo. (1996). *Quick Changeover for Operators: SMED System (Shop Floor)*. New York: Productivity Press.

Stolovitch, Harold D. and Keeps, Erica J. (2002). *Telling Ain't Training*. Alexandria, Virginia: American Society for Training & Development.

Sullivan, William G., Wicks, Elin M. and Luxhoj, James T. (2003). *Engineering Economy, 12th edition*. Upper Saddle River, New Jersey: Prentice Hall, Pearson Education, Inc.

Tague, Nancy. (2005). *The Quality Toolbox, 2nd edition*. Milwaukee, Wisconsin: Quality Press.

Wortman, Bill. (2001). *The Certified Six Sigma Black Belt Primer, 1st edition*. West Terre Haute, Indiana: Quality Council of Indiana.

Womack, James and Jones, Daniel T. (1996). *Lean Thinking*. New York, New York: Simon & Schuster.

Notes

www.ingramcontent.com/pod-product-compliance
Lightning Source LLC
Chambersburg PA
CBHW041726210326
41598CB00008B/794